Mathematics Research Developments Series

A Compendium of Univariate Statistical Distributions
Mohammad Ahsanullah, PhD
and Mohammad Shakil, PhD
ISBN: 979-8-89530-727-4
eBook ISBN: 979-8-89530-904-9

The Direct Integration of Ordinary Differential Equations
A.B Urdaletova, PhD
and S.K. Kydyraliev, PhD
eBook ISBN: 979-8-89530-690-1

Mathematics for Optimal Decision-Making and Applications: A Deterministic Approach
Rami A. Maher, PhD
ISBN: 979-8-89530-560-7
eBook ISBN: 979-8-89530-763-2

Numerical Optimization on Time Scales
Svetlin Georgiev
ISBN: 979-8-89530-466-2
eBook ISBN: 979-8-89530-503-4

The Laplace Transform and Its Applications in Science
Victor Martinez-Luaces, PhD (Editor)
ISBN: 979-8-89113-747-9
eBook ISBN: 979-8-89530-293-4

Selected Topics in Fractional Differential Equations
Zahia Drici, PhD,
J. Vasundhara Devi, PhD
and Farzana A. McRae, PhD (Editors)
ISBN: 979-8-89530-301-6
eBook ISBN: 979-8-89530-428-0

More information about this series can be found at
https://novapublishers.com/product-category/series/mathematics-research-developments/

Diophantine Equations

nova
science publishers
www.novapublishers.com

Vladimír Medvid'

Diophantine Equations

nova
science publishers
www.novapublishers.com

DOI: https://doi.org/10.52305/VFKF1121

Library of Congress Cataloging-in-Publication Data

ISBN: 979-8-89530-955-1 (Hardcover)
ISBN: 979-8-90134-020-2 (eBook)

Published by Nova Science Publishers, Inc. † New York

Contents

Preface

A Diophantine equation (sometimes also Diophantine) in mathematics is an indeterminate polynomial equation that allows the variables to take on values only from the range of integers. Diophantine problems have fewer equations than unknown variables and involve finding integers that are solutions to all the equations of the system. In more technical terms, they define an algebraic curve, an algebraic surface, or a more general shape, and look for grid points on it.

The word Diophantine refers to the 3rd-century ancient mathematician Diophantus of Alexandria, Egypt, who studied such equations and was also one of the first mathematicians to introduce symbolism in algebra. The mathematical study of Diophantine problems, initiated by Diophantus, is now known as "Diophantine analysis." A linear Diophantine equation is an equation of two sums of monomials of the first or zeroth order.

While the individual equations represent a kind of puzzle and have been investigated many times, the formulation of the general theory of Diophantine equations was only obtained in the twentieth century, later than the theory of quadratic forms.

Related to the solution of Diophantine equations is Hilbert's 10th problem, which asked for the existence of an algorithm that can decide whether there is a solution for any Diophantine equation. The possibility of the existence of such an algorithm was ruled out by Matijasevich's theorem in 1970. Yuri Vladimirovich Matijasevich demonstrated that a decision algorithm cannot be found for equations with more than nine variables. However, Wiles's method of proving Fermat's Big Theorem suggests that there should be a decision algorithm for Diophantine problems with three variables. The question still remains unanswered: What is the lowest number of variables for which the existence of a solution to the Diophantine equation is undecidable?

In this book, we present several types of Diophantine equations from simple to more complex. The aim of this work is to interest readers at different levels of knowledge and education. Therefore, this book is suitable for gifted high school students and university students.

The text also contains word problems, which are tempting for readers who enjoy such problems.

The book consists of six chapters.

In the first chapter, the authors present the Euclidean algorithm and an extended Euclidean algorithm for linear Diophantine equations with two unknowns.

The second chapter addresses Diophantine linear equations with three unknowns and examines solvability conditions.

The third chapter focuses on Diophantine equations of the second degree.

Special Quadratic Diophantine equation of the second degree with two unknowns, Diophantine bilinear equation, Pythagorean equation, Algorithms of solving The Diophantine equation, and Diophantine equation $x^2 + y^2 = \alpha . z^2, \alpha \in \mathrm{N}$.

Linear Diophantine equations of n unknowns are discussed in the fourth chapter. Two methods: Method of solving congruences and the Method of reduction to a smaller number of unknowns.

The Pythagorean trinity is introduced in Chapter 5.

In the sixth chapter, Fermat's equation, that is, the Diophantine equation of degree above, is presented, and its solution is discussed.

In the seventh chapter, linear Diophantine equations are solved similarly to those in the first chapter, but they are presented within a more rigorous professional framework through definitions, theorems with proofs, and algorithms. They are also presented with examples and exercises, along with the corresponding solutions. In addition, problems of Diophantine equations in Z_n spaces are solved here.

The problems mentioned, along with their study, are suitable for students of mathematics at universities and also for scientists in the relevant field.

Finally, in the eighth chapter, systems of linear Diophantine congruences and systems of linear Diophantine equations are solved.

Similarly to the previous chapter, the problems listed, along with their study, are suitable for students of mathematics at universities and also for scientists in the relevant field.

Vladimír Medviď,
Dr., University of Žilina, Slovakia

Chapter 1

Linear Diophantine Equations with Two Unknowns

Abstract

Diophantine equations are a fascinating mathematical theory in the field of algebra and number theory. They are an eternal inspiration for students of all ages, from elementary school to university, in mathematics.
The main goal of this chapter is to show the methodology of solving linear Diophantine equations with n unknowns, especially for students of the mentioned categories.
In this chapter, we propose two methods for solving these tasks subsequently

- Method of solving congruences
- Method of reduction to a smaller number of unknowns

We elaborated these methods theoretically in detail and then gave solutions based on concrete examples. At the end, we also provided the reader with unsolved tasks for self-practice.
Method of solving congruences: When calculating specific examples, we will proceed similarly to the previous proof, that is, we will solve the congruence according to the modulus, which is the greatest common divisor of n - 1 coefficients, provided that it is different from 1, then we will solve the congruence with one unknown and subsequently we gradually calculate the remaining variables.
Method of reduction to a smaller number of unknowns: The supervisor proposed the following procedure. The advantage of this method lies in the fact that we will not need congruences; we will only need the extended Euclidean algorithm.
Educational goal. These methods were presented at selection seminars to talented high school students and university students studying mathematics, and then tests were conducted to evaluate their ability to solve specific tasks related to the mentioned issue.
The students perfectly managed almost all assigned tasks with excellent results.
The aforementioned methods for solving the tasks were also studied with interest by secondary school and university teachers.

With respect to the educational goals, this chapter is recommended for students in mathematical classes and students of mathematics with a specialization in algebra or number theory.

Keywords: linear Diophantine equation, greatest common divisor, congruence, reduction, positive integer, extended Euclidean algorithm

1.1. Introduction

In this chapter, we will deal with a frequently occurring type of Diophantine equation, which is associated with word problems. Therefore, we will address it in detail and present various solution methods.

Definition 1.1. We call each linear Diophantine equation with two unknowns an equation in the form

$$a\,x + by = c\,, \tag{1.1}$$

where x, y are unknown, a, b, c are integers, $a \neq 0, b \neq 0$, see e.g., Mordell [5] or Schmidt and Wolfgang [7].

We can divide the solution of equation (1.1) into four cases:

$(I)\ a > 0, b > 0$

$(II)\ a > 0, b < 0$

$(III)\ a < 0, b > 0$

$(IV)\ a < 0, b < 0$

Let us realize that it is enough to solve only case (I), because if we solve (IV), then it is enough to multiply equation (3.1) by the number -1 and this will convert to (I). Problems (II) and (III) are symmetric, so we only need to solve (II), for example. If we consider case (II), then by substituting $y = -z$, we are again converted to case (I), where we solve the equation for the unknowns x, z with positive coefficients and at the end it is enough to return to substitution. From now on, we consider equation (1.1) for $a > 0, b > 0$.

Definition 1.2. Let a, b be positive integers. We will say that the positive integer δ is the greatest common divisor of numbers a, b if and only if:

(I) $\delta | a \wedge \delta | b$,

(II) for all positive integers e: $(e | a \wedge e | b) \Rightarrow e | \delta$

We will write down the fact that δ is the greatest common divisor of the numbers a, b

$$\delta = gcd(a, b).$$

If only (I) holds, we call the number δ the common divisor of the numbers a, b.

1.2. Euclidean Sequencing Algorithm

We use the Euclidean sequencing algorithm to determine the largest common divisor. Let $a, b, a > b$ be any given integers. The Euclidean sequencing algorithm can be described as follows: See Medvid [4].

Consider the equation

$$ax + by = 1. \tag{1.2}$$

Let us perform a variation of the Euclidean algorithm

$$a = bq_1 + r_1 \tag{1.3}$$

$$b = q_2 r_1 + r_2 \tag{1.4}$$

$$r_1 = q_3 r_2 + r_3 \tag{1.5}$$

$$\dots$$

$$r_{k-2} = q_k r_{k-1} + r_k \tag{1.6}$$

$$r_{k-1} = q_{k+1} r_k + 0, r_k = gcd(a, b) \tag{1.7}$$

The greatest common divisor of the numbers a, b is the last nonzero remainder r_k.

Continue this procedure all the way back to the top.

From the above, it follows that

$$r_1 = a - bq_1,$$

$$r_2 = b - q_2(a - bq_1) = -q_2 a + (1 + q_1 q_2)b, \ldots,$$

$$r_k = gcd(a, b) = \alpha a + \beta b,$$

where α and β are integer numbers. If a, b are coprime numbers, i.e., $gcd(a, b) = 1$, then the equation $a\alpha + b\beta = 1$ is obviously satisfied for α, β integers.

The reversed opposite implication is also valid: If there are α, β integer numbers such that $a\alpha + b\beta = 1$ then a, b are coprime numbers, i.e., $gcd(a, b) = 1$.

If equation (1.1) has an integer solution (x_0, y_0), then $\delta = gcd(a, b)$ is a divisor of c.

The opposite implication is also valid. More precisely, if $\delta = gcd(a, b)$ is a divisor of c then equation (1.1) has an integer solution (x_0, y_0). If $\delta = gcd(a, b)$ then $\frac{a}{\delta}, \frac{b}{\delta}$ are coprime numbers, i.e., $d\left(\frac{a}{\delta}, \frac{b}{\delta}\right) = 1$. Accordingly, there exists α, β integer numbers, such that

$$\frac{a}{\delta}\alpha + \frac{b}{\delta}\beta = 1 \qquad (1.7^*)$$

Multiplying $(1.7 *)$ by c yields $a\frac{\alpha c}{\delta} + b\frac{\beta c}{\delta} = c$. Consequently, the integer numbers $x_0 = \frac{\alpha c}{\delta}, y_0 = \frac{\beta c}{\delta}$ are the solutions of equation (1.1).

The sentence applies:

Theorem 1.1. Equation (1.1) has an integer solution if and only if the greatest common divisor of the numbers a, b is also a divisor of the number c, t. j. $\delta = gcd(a,b), \delta \mid c$.

Proof.

Since the theorem is stated in the equivalence form, we will have to prove two things:

(I) If equation (1.1)has an integer solution, then $\delta \mid c$.

(II) If $\delta \mid c$, then equation (1.1) has integer solutions.

ad(I) Let x_0, y_0 denote the integer solution of equation (3.1); then it applies

$$ax_0 + by_0 + c = 0.$$

Since $\delta = gcd(a, b)$, there exist integers a_1, b_1 such that

$$a = a_1\delta, b = b_1\delta.$$

Then

$$a_1 d\delta + b_1\delta y_0 + c = 0$$

$$d\delta\,(-a_1 x_0 - b_1 y_0) = c.$$

From there, he must $\delta \mid c$, which we wanted to prove.

ad(II) Since $\delta \mid c$, there exists an integer c_1 such that $c = c_1\delta$. Let's take the coefficients a, b from equation (1.1) and let's apply the extended Euclidean algorithm to them. The one for us finds integers α, β such that the following Bezout equality is satisfied

$$a\alpha + b\beta - \delta = 0.$$

Now we multiply the obtained equality by the number c1 and get

$$a\alpha c_1 + b\beta c_1 - \delta c_1 = 0$$

$$a\alpha c_1 + b\beta c_1 - c = 0$$

$$a(-\alpha c_1) + b(-\beta c_1) + c = 0.$$

Comparing with equation (1.1), we obtain

$$x_0 = -\alpha c_1 \, y_0 = -\beta c_1$$

and that is some integer solution we should have found. This completes the proof.

Theorem 1.1 has a very interesting consequence. If $\delta = gcd(a, b) = 1$, it will have equation (1.1) always has an integer solution, since it is always true that $1 \mid c$ for all c. Moreover, every integrally solvable equation (according to Theorem 3.1, this means that $\delta \mid c$) can be divided by the number δ and thus convert to the case where we already have $gcd(a, b) = 1$.

Note: If equation (1.1) is solvable, then it does not have a single solution, but infinitely many solutions. If we determine one of them in any way (using Euclid's algorithm or by experiment), we will get all the others very easily.

Now let us pay attention to equation (1.2) and the subsequent Euclidean algorithm. The equation has an integer solution if only if $r_k = d(a, b) = 1$, where r_k is the last nonzero remainder, i.e., equations (1.6) and (1.7) can be rewritten as follows:

$$r_{k-2} = q_k r_{k-1} + 1 \qquad (1.8)$$

$$r_{k-1} = q_{k+1} . 1 + 0, 1 = gcd(a, b) \qquad (1.9)$$

Starting from the first equation (1.8) we get

$$1 = r_{k-2} - q_k r_{k-1} \qquad (1.10)$$

$$r_{k-1} = r_{k-3} - q_{k-1} r_{k-2}, \qquad (1.11)$$

Thus

$$1 = r_{k-2} - q_k(r_{k-3} - q_{k-1} r_{k-2}) = -q_k r_{k-3} + (1 + q_{k-1} q_k) r_{k-2} \qquad (1.12)$$

Continue this procedure all the way back to the top.

Example: Solve the Diophantine equation

$$30031.x + 2311.y = 1. \tag{1.13}$$

Solution. The first step is to find the largest common divisor of the numbers 30031 and 2311.

Both coefficients of (1.13) are designed according to a well-known method so that they are prime numbers. Indeed, $2.3.5.7.11.13 + 1 = 30031$ and $2.3.5.7.11 + 1 = 2311$.

That means that the largest common divisor of both numbers is $gcd(a, b) = 1$. This is a necessary and sufficient condition for our equation to have an integer solution.

Now, we can apply the above-mentioned algorithm to find a solution to our equation.

$$\begin{array}{rcl|rcl}
30031 &=& 2311.12 + 2299 & 1 &=& -963.3011 + 12514.2311 \\
2311 &=& 2299.1 + 12 & 1 &=& 958.2311 - 9963.2299 \\
2299 &=& 12.191 + 7 & 1 &=& -5.2299 + 958.12 \\
12 &=& 7.1 + 5 & 1 &=& 3.12 - 5.7 \\
7 &=& 5.1 + 2 & 1 &=& -2.7 + 3.5 \\
5 &=& 2.2 + 1 & 1 &=& 1.5 - 2.2 \\
2 &=& 1.2 + 0 \downarrow & 1 &=& 0.2 + 1.1 \uparrow
\end{array} \tag{1.14}$$

The solution is therefore $x = -963, y = 12514$.

Remainders are $r_1 = 2299$, $r_2 = 12$, $r_3 = 7$, $r_4 = 5$, $r_5 = 2$ and $r_6 = 1$.

It is worthwhile to mention one interesting fact in the reversed procedure. Coefficients by the r_{i-1} and r_{i+1} remainders are the same, namely

1.1 and 1.5,
−2.2 and −2.7,
3.5 and 3.12,
−5.7 and −5.2299,
958.12 and 958.2311,
−963.2299 and −963.3011.

Generally,

$$1 = -a_{i+1}r_i + a_i r_{i+1} \tag{1.15}$$

$$r_{i+1} = r_{i-1} - q_{i+1}r_i. \tag{1.16}$$

Substituting the right side of equation (1.16) instead of r_{i+1} into equation (1.15) we obtain

$$1 = a_i r_{i-1} - (a_{i+1} + a_i q_{i+1}) r_i. \tag{1.17}$$

Thus, r_{i-1} and r_{i+1} remainders have the same coefficient a_i and

$$a_{i-1} - (a_{i+1} + a_i q_{i+1}). \tag{1.18}$$

For this reason, the above procedure can be simplified by one entry and alternate signs

$$\begin{array}{l|l}
30031 = 2311.12 + 2299 & (-)963.12 + 958 = 12514 \\
2311 = 2299.1 + 12 & (+)958.1 + 5 = 963 \\
2299 = 12.191 + 7 & (-)\ 5.191 + 3 = 958 \\
12 = 7.1 + 5 & (+)\ 3.1 + 2 = 5 \\
7 = 5.1 + 2 & (-)\ 2.1 + 1 = 3 \\
5 = 2.2 + 1 & (\mp)\ 1.2 + 0 = 2 \\
2 = 1.2 + 0 \downarrow & (-)\ 0.1 + 1 = 1 \uparrow
\end{array} \tag{1.19}$$

The solution is the same as the solution obtained by the original procedure, which is

$$x = -963\ y = 12514.$$

This simplified procedure is particularly suitable for constructing a simple and efficient algorithm. The algorithm using this simplified procedure is much faster than the original algorithm and can solve the Diophantine equations with relatively large coefficients in a "reasonable," polynomial time.

Let us turn our attention to equation (1.1). If this equation is solvable, then it does not have a single solution but infinitely many solutions. If we determine one of them in any way (using the Euclidean algorithm or experimentally), then we will obtain all solutions easily.

Let (x_0, y_0) be an integer solution of equation (1.1), i.e.,

$$ax_0 + by_0 = c. \tag{1.20}$$

By subtracting equation (1.20) from (1.1), we obtain

$$a(x - x_0) + b(y - y_0) = 0 \quad (1.21)$$

Dividing by the largest common divisor $\delta = gcd(a, b)$yields

$$\frac{a}{\delta}(x - x_0) = -\frac{b}{\delta}(y - y_0) \quad (1.22)$$

The numbers $\frac{a}{\delta}$ and $\frac{b}{\delta}$ are the coprime numbers. Thus, from equation (1.22) it follows that $\frac{a}{\delta} \mid (y - y_0)$. There is an integer number t such that $(y - y_0) = \frac{a}{\delta}.t$.

Substituting into equation (1.22) and dividing by $\frac{a}{\delta}$ we obtain $(x - x_0) = -\frac{b}{\delta}t$. Consequently, all integer solutions of the equation (1.1) can be expressed in the form

$x = x_0 - \frac{b}{\delta}t$ and $y = y_0 + \frac{a}{\delta}t$, where t is an arbitrary integer number. See e.g., Shallit [6] or Weinberger [9].

Examples

Example 1. Solve the Diophantine equation

$$10x + 4y = 16. \quad (1.23)$$

Solution. $a = 10, b = 4, c = 16, \quad gcd(10,4) = 2, and\ 2 \mid 16$. This means that the equation has an integer solution. It is easy to see that, for example $x_0 = 2, y_0 = -1$ is the one simple solution, i.e.,

$$10x_0 + 4y_0 = 16. \quad (1.24)$$

Subtracting equation (1.24) from (1.23) we obtain

$$10(x - x_0) + 4(y - y_0) = 0 \quad (1.25)$$

After dividing by the largest common divisor $gcd(10,4) = 2$ we obtain

$$5(x - x_0) = -2(y - y_0). \quad (1.26)$$

The numbers 5 and -2 are the coprime numbers. It means that

$5 \mid (y - y_0)$ and $-2 \mid (x - x_0)$.

Clearly, there is an integer number t, such that $(y - y_0) = 5t$ and $(x - x_0) = -2.t$.

Which follows

$x = x_0 - 2t$ and $y = y_0 + 5t$. Substituting $x_0 = 2, y_0 = -1$ we obtain that all integer solutions are

$$x = 2 - 2t$$
$$y = -1 + 5t,$$

where t is an arbitrary integer number.

Example 2. Find out which of the following equations are solvable in the Z set of integer numbers

$$3x + 2y = 1$$

$$4x - 6y = 8$$

12x+18y=30

$$15x - 9y = 14.$$

Solution.

a) $a = 3, b = 2, c = 1.$ $gcd(3,2) = 1, and\ 1 \mid 1$. This means that the equation has an integer solution.
b) $a = 4, b = -6, c = 8.$ $d(4, -6) = 2, and\ 2 \mid 8$. Thus, the equation has an integer solution.
c) $a = 12, b = 18, c = 30.$ $gcd(12,18) = 6, and\ 6 \mid 30$. Consequently, the equation has an integer solution.

d) $a = 15, b = -9, c = 14, gcd(15, -9) = 3, and\ 3 \nmid 14$. This means that the equation has no integer solution. $3 \nmid 14$ means that the number 3 does not divide the number 14 in the Z set of integer numbers.

Example 3. How many ways can we cut a strip 50 cm long into 6 cm pieces?

Solution. Let the letter a represent the ribbon 6 cm long, the letter b represents the ribbon 4 cm long and the letter c represents the whole strip 50 cm long.

The problem can be formulated by the Diophantine equation

$$6x + 4y = 50, \tag{1.27}$$

$$x \geq 0\,, y \geq 0. \tag{1.28}$$

It is easy to verify that, for example, $x_0 = 7, y_0 = 2$ is one simple solution, since

$$6x_0 + 4y_0 = 50 \tag{1.29}$$

By subtracting the equation $(e.3.3)$ from $(e.3.1)$ we obtain

$$6(x - x_0) + 4(y - y_0) = 0 \tag{1.30}$$

After dividing by the largest common divisor $gcd(6, -4) = 2$ we get

$$3(x - x_0) = -2(y - y_0). \tag{1.31}$$

The numbers 3 and -2 are the coprime numbers. Consequently,

$3 \mid (y - y_0)$ and $-2 \mid (x - x_0)$.

There is an integer number t, such that

$(y - y_0) = 3t$ and $(x - x_0) = -2t$.

Thus,

$x = x_0 - 2t$ and $y = y_0 + 3t$. Substituting $x_0 = 7, y_0 = 2$ we obtain that all integer solutions have the form

$x = 7 - 2t$ y =2+3t,

where t is an arbitrary integer number.

Given the conditions $(e.3.2)$ for non-negativity of solutions we have

$7 - 2.t \geq 0$ and $2 + 3t \geq 0$, i.e.,

$-\frac{2}{3} \leq t \leq \frac{7}{2}$ and $t \in Z$.

For $t \in \{0, 1, 2, 3\}$ we obtain all solutions $\{(\mathbf{7}, \mathbf{2}), (\mathbf{5}, \mathbf{5}), (\mathbf{3}, \mathbf{8}), (\mathbf{1}, \mathbf{11})\}$.

The above examples were solved easily due to their small coefficients. Indeed, it was enough to test some numbers and verify that the equation holds.

Perhaps it will not be so easy to solve the equation when the coefficients are large numbers. Then it will not be possible to find any initial solution.

Now, the question is, how to find a solution in the given case.

One known method, based on the division of numbers, will be very helpful here.

Thus

if $a \mid p(x)$ then $a \mid \big(p(x) + kaq(x)\big)$ where $a, k \in Z$.

By adding or subtracting a suitable multiple of the number a to $p(x)$ we reduce $p(x)$.

We will show the use of this method in the following example.

Example 4. Solve the Diophantine equation $-17x + 91y = 24$.

Solution. $a = -17, b = 91, c = 24, d(-17,91) = 1$ *and* $1 \mid 24$, i.e., the equation has an integer solution.

We express the term $17.x$ from the equation $-17x + 91y = 24$, i.e.,

$$17x = 91y - 24. \quad (1.32)$$

It follows from the above equation that $17 \mid (91.y - 24)$. If we add to every summand of the expression $91.y - 24$ a suitable multiple of the number 17 then we get the expression divisible by the number 17. We choose the added multiples so that the absolute value of each of the summands decreases as much as possible. Therefore, we add to the first member $-5.17.y$ and to the second member 1.17. We get $6.y - 7$ and this expression is divisible by 17. There exists $u \in Z$ such that it is fulfilled $6.y - 7 = 17.u$. Thus

$$6y = 17u + 7 \quad (1.33)$$

The expression $17.u + 7$ is divisible by 6, i.e., $6 \mid 17u + 7$. Therefore, again, when adding a suitable multiple of 6 to each of the sums, we get the expression divisible by 6. There exists $v \in Z$, such that it holds $6v = (17u + 7\,) + (-18u - 6) = 1 - u \Rightarrow u = 1 - 6v$.

Substituting it into the right-hand side of equation $(e.4.2)$, we get

$$6y = 17(1 - 6v) + 7 = 24 - 6.17v.$$

Now dividing by 6 yieds $y = 4 - 17v$.

After substitution into the left-hand side of equation $(e.4.1)$ we get

$$17x = 91(4 - 17v) - 24 = 340 - 17.91v.$$

Consequently, $x = 20 - 91v$.

Finally, we get the solutions $x = 20 - 91v, y = 4 - 17v,\ v \in Z$.

Proof of the correctness: $-17(20 - 91v) + 91(4 - 17v) = 24$.

Exercises

1. Find all integer solutions to the equation:

$a)\ 2x + 3y = 7, b)\ 4x + 5y = 9, c)\ 7x - 2y = 5,$

$d)\ 3x + 4y = 15, e)\ 5x + 8y = 23, f)\ 4x - 7y = 13,$
$g)\ 6x + 7y = 20, h)\ 11x + 5y = 17,$ i) $9x - 42y = 15,$
$j) - 17x + 91y = 24\ k)\ 30x - 91y = 2, l)\ 16x + 21y = 200$

2. A group of less than 100 trainees entered the seven steps, but the last row was incomplete - two trainees were missing. When this group entered the nine steps, there were two extra practitioners. How many trainees were there in the group?
3. The ages of a woman and a man are "inverted" numbers, and their difference is a fifth of the woman's age. Determine the age of both. (By reversed numbers we mean, for example, the numbers 1234 and 4321).
4. Nineteen points in the plane, no three of which lie on the same straight line, are connected to form triangles and quadrilaterals. How many triangles and how many quadrilaterals can we draw if each point is the vertex of a single image?
5. The transport company has two types of sets available to transport 12,000 people: set A can accommodate 90 people, and set B can accommodate 150 people. How many of each set will be used to keep the total number to a minimum?
6. In the parliament of a certain state, 253 deputies belong to three parties: government, neutral, and opposition. The ruling party is the strongest, but does not have an absolute majority. To achieve it, he needs exactly 28% of the votes of the neutral party. Both non-opposition parties together have a majority of more than two-thirds. How many MPs does each party have in parliament?

Solutions

1. $a)$ Let's try and easily find that one solution of the equation is

$(d(2,3) = 1)$:
$2x + 3y = 7,$
$x_0 = 2, y_0 = 1,$

because $2.2 + 3.1 = 7$.

By subtracting the second equation from the first, we get:

$$2(x-2)+3(y-1)=0$$

from where

$$2(x-2)=-3(y-1)$$
$$\Rightarrow x-2=3m, y-1=-2m, m \in Z.$$

All solutions (integer) of the given equation are:

$$x=2+3m, y=1-2m, m \in Z.$$

Correctness test: $2(2+3m)+3(1-2m)=7$.

$b)$ $d(4,5)=1$:
$4x+5y=9$,
$x_0=2$,
$y_0=1$
$\Rightarrow 4(x-1)+5(y-1)=0$
$\Rightarrow 4(x-1)=-5(y-1)$
$\Rightarrow x-1=5m, y-1=-4m, m \in Z$
$\Rightarrow x=1+5m, y=1-4m, m \in Z.$

Correctness test: $4(1+5m)+5(1-4m)=9$.

$c)$ $d(7,-2)=1$:
$7x-2y=5$,
$x_0=1, y_0=1$
$\Rightarrow 7(x-1)-2(y-1)=0$
$\Rightarrow 7(x-1)=2(y-1)$
$\Rightarrow x-1=2m, y-1=7m, m \in Z$
$\Rightarrow x=1+2m, y=1+7m, m \in Z.$

Correctness test: $7(1+2m)-2(1+7m)=5$.

$d)$ $d(3,4)=1$:
$3x+4y=15, x_0=1, y_0=3$

$\Rightarrow 3(x-1)+4(y-3)=0$
$\Rightarrow 3(x-1)=-4(y-3) \Rightarrow x-1=4m, y-3=-3m, m \in Z$
$\Rightarrow x=1+4m, y=3-3m, m \in Z.$

Correctness test: $3(1+4m)+4(3-3m)=15.$

$e)$ $d(5,8)=1$:
$5x+8y=23,$
$x_0=3, y_0=1$
$\Rightarrow 5(x-3)+8(y-1)=0$
$\Rightarrow 5(x-3)=-8(y-1)$
$\Rightarrow x-3=8m, y-1=-5m, m \in Z$
$\Rightarrow x=3+8m, y=1-5m, m \in Z.$

Correctness test: $5(3+8m)+8(1-5m)=23.$

$f)$ $d(4,-7)=1$:
$4x-7y=13,$
$x_0=5, y_0=1$
$\Rightarrow 4(x-5)-7(y-1)=0$
$\Rightarrow 4(x-5)=7(y-1)$
$\Rightarrow x-5=7m, y-1=4m, m \in Z$
$\Rightarrow x=5+7m, y=1+4m, m \in Z.$

Correctness test: $4(5+7m)-7(1+4m)=13.$

$g)$ $d(6,7)=1$:
$6x+7y=20,$
$x_0=1, y_0=2$
$\Rightarrow 6(x-1)+7(y-2)=0$
$\Rightarrow 6(x-1)=-7(y-2)$
$\Rightarrow x-1=7m, y-2=-6m, m \in Z$
$\Rightarrow x=1+7m, y=2-6m, m \in Z.$

Correctness test: $6(1+7m)+7(2-6m)=20.$

$h)$ $d(11,5)=1$:
$11x+5y=17,$

$x_0 = 2, y_0 = -1$
$\Rightarrow 11(x-2) + 5(y+1) = 0$
$\Rightarrow 11(x-2) = -5(y+1)$
$\Rightarrow x - 2 = 5m, y + 1 = -11m, m \in Z$
$\Rightarrow x = 2 + 5m, y = -1 - 11m, m \in Z.$

Correctness test: $11(2 + 5m) + 5(-1 - 11m) = 17.$

$i)$ $d(9, -42) = 3$:
$3/15$: $9x - 42y = 15.$

After multiplying the given equation by $\frac{1}{3}$ we get

$3 - 14y = 5,$
$x_0 = 11, y_0 = 2$
$\Rightarrow 3(x - 11) - 14(y - 2) = 0$
$\Rightarrow 3(x - 11) = 14(y - 2)$
$\Rightarrow x - 11 = 14m, y - 2 = 3m, m \in Z$
$\Rightarrow x = 11 + 14m, y = 2 \mp 3m, m \in Z.$

Correctness test: $9(11 + 14m) - 42(2 + 3m) = 15.$

$j)$ $d(-17,91) = 1$:

$-17x + 91y = 24$
$\Rightarrow 17x = 91y - 24$
$\Rightarrow 17/(91y - 24).$

If we add a suitable multiple of the number 17 to each of the addends of the expression $91y - 24$ we get an expression divisible by the number 17. We choose the added multiples so that the absolute value of each addend is reduced as much as possible. So, let's add $-5.17y$to the first term and 1.17 to the second. we get $6y - 7$, which is an expression divisible by 17. So there exists u, $u \in Z$ such that: $6y - 7 = 17u$ from where

$6y = 17u + 7.$

The expression $17u + 7$ is divisible by the number 6. Therefore, after adding a suitable multiple of the number 6 to each of the addends, we again get an expression divisible by the number 6. Therefore, $v \in Z$ exists such that:

$$6y = (17u + 7) + (-18u - 6) = 1 - u \rightarrow u = 1 - 6v..$$

If we substitute it into the expression on the right side of equation (5.7), we get:

$6y = 17(1 - 6v) + 7 = 24 - 6.17v$, from where, after multiplying by $\frac{1}{6}$ we get: $y = 4 - 17v$. After substituting for y in (5.6) we have:

$$17x = 91(4 - 17v) - 24 = 340 - 17.91v,$$

whence $x = \ 20 - \ 91v$.

We thus got all the solutions of the given equation:

$$x = \ -91v + 20, y - 4 - 17v, v \ \in Z.$$

Correctness test: $-17(-91v + 20) + 91(4 - 17v) = 24$.

Note: In solving this task, we used the method used in solving the Diophantine equation in the case when it is not easy to determine some of its solutions by experiment.

$k)\ d(36, -49) = 1$:
$36x - 49y = 2$
$\Rightarrow 36x = 2 + 49y$
$\Rightarrow 36/(2 + 49y)$.

$\exists\ u \in Z$ such that

$36u = (2 + 49y) + (36 - 36y) = 38 + 13y \ \Rightarrow 13y = 36u - 38$
$\Rightarrow 13/(36u - 38)$
$\exists v \in Z$ such that
$13v = (36u - 38) + (-39u + 39) = 1 - 3u$
$\Rightarrow 3u = 1 - 3v$

$\Rightarrow 3/(1-3v)$
$\Rightarrow \exists m \in Z$

such that

$3m = (1-13v)+12v = 1-v$
$\Rightarrow v = 1-3m$
$\Rightarrow 3u = 1-13(1-3m) = -12+3.13u$
$\Rightarrow u = -4+13m$
$\Rightarrow 13y = 36(-4+13m)-38 = 13.36m-128$
$\Rightarrow y = 36m-14$
$\Rightarrow 36x = 2+49(36m-14) = 36.49m-684$
$\Rightarrow x = 49m-19.$

We thus obtained all integer solutions of the given equation:

$x = 49m-19, y = 36m-14, m \in Z.$

Correctness test: $36(49m-19)-49(36m-14) = 2.$

l) $d(16,21) = 1$:
$16x+21y = 200,$
$x_0 = 2, y_0 = 8$
$\Rightarrow 16(x-2)+21(y-8) = 0$
$\Rightarrow x-2 = 21m$
$\Rightarrow x = 2+21m, y-8 = -16m$
$\Rightarrow y = 8-16m, m \in Z.$

Correctness test: $16(2+21m)+21(8-16m) = 200.$

2. If we denote the number of all seven steps by x, the number of trainees is $7x-2$, and if we denote the number of nine steps by y, there will be $9y+2$ trainees. Obviously, $7x-2 = 9y+2$ must be $7x-9y = 4$, which is a Diophantine equation with by two unknowns, which has a solution because $d(7,-9)-1$ and $1/4$.

It is easy to see that one solution to this equation is $x_0 = 7, y_0 = 5$. Therefore $7(x-7)-9(y-5) = 0 \leftrightarrow 7(x-7) = 9(y-5)$ from where $x-7 = 9m, y-5 = 7m, m \in Z.$

However, from the fact that it must be $0 < 7x - 2 = 9y + 2 < 100$, we get: $0 < 7(7 + 9m) - 2 < 100$, i.e., $0 < 47 + 63m < 100$ or $0 < 9(5 + 7m) + 2 < 100 \leftrightarrow 0 < 47 + 63m < 100$.

However, the only solution to this system of inequalities in the domain of whole numbers is $m = 0$m=0, which means that there were 47 trainees in the group.

3. Let's mark $10x + y$ as the man's age. Then the woman's age will be $10y + x$. It is obvious that x, y are single-digit positive integers. From the fact that the age difference between the two is a fifth of the woman's age, we conclude:

$$10(x-y)y - x = \frac{1}{5}(10y + x)$$
$$\Rightarrow 9(x-y) = \frac{1}{5}(10y + x)$$
$$\Rightarrow 45x - 45y = 10y + x$$
$$\Rightarrow 44x - 55y = 0,$$
$$x_0 = 5, y_0 = 4$$
$$\Rightarrow 4(x-5) - 5(y-4) = 0$$
$$\Leftrightarrow 4(x-5) = 5(y-4)$$
$$\Rightarrow x - 5 = 5m, y - 4 = 4m.$$

All solutions of the given equation are:

$$x = 5 + 5m, y = 4 + 4m, m \in Z.$$

We can easily verify that we get a positive one-digit solution only for $m = 0$. It follows that the man is 54 years old, and the woman is 45 years old.

With the test, we can easily justify the correctness of the solution to the task.

4. Let x denote the number of triangles and y the number of quadrilaterals. The following applies to the number of vertices of these shapes: $3x + 4y = 19$. Since $d(3,4) = 1$, the task has a solution. One solution is obviously $x_0 = 1, y_0 = 4$.

Therefore $3(x-1) + 4(y-4) = 0$. Since $3(x-1) = -4(y-4)$, there exists $m \in Z$ such that $x - 1 = 4m, y - 4 = -3m$. All solutions of the given equation g are therefore:

$$x = 1 + 4m, y = 4 - 3m, m \in Z.$$

Apparently, only non-negative solutions of the equation are suitable for the given task.

$$x \geq 0 \Leftrightarrow 1 + 4m \geq 0 \Leftrightarrow m \geq -\frac{1}{4}, y \geq 0 \Leftrightarrow 4 - 3m \geq 0 \Leftrightarrow m \leq \frac{4}{3}.$$

They meet these conditions

$m = 0 \Rightarrow x = 1, y = 4$ and $m = 1 \Rightarrow x = 5, y = 1$

This task has two solutions:

a) 1 triangle and 4 quadrilaterals
b) 5 triangles and 1 quadrilateral.

5. Let x denote the number of sets A, y the number of sets B. It follows from the conditions of the task that $90x + 150y = 12000$.. Since $d(90{,}150) = 30$ and $30/12000$, the given equation has an integer solution. However, we are only interested in an integer solution such that $x + y$ is minimal. After dividing the given equation by the greatest common divisor of the coefficients, i.e., j, with the number 30, we get: $3x + 5y = 400$. It follows that $3x = 400 - 5y \Rightarrow 3/(400 - 5y)$.

So there exists $m \in Z$ such that

$$3m = 400 - 5y + (-399 + 6y) = 1 + y \Rightarrow y = -1 + 3m, m \in Z..$$

But then $3x = 400 - 5(-1 + 3m) = 405 - 15m \rightarrow x = 135 - 5m$.. It follows from the non-negativity of the sought solution that $x \geq 0 \Leftrightarrow 135 - 5m \geq 0 \Leftrightarrow m \leq 27$ and $y \geq 0 \Leftrightarrow -1 + 3m \geq 0 \Leftrightarrow m \geq \frac{1}{3}$. The minimum sum condition gives us $x + y = 134 - 2m$, which will be minimum for the whole $m \in \langle\frac{1}{3}, 27\rangle$ if and only if $m = 27$. It follows that $x - 0, y = 80$.

Conclusion: 80 sets of type B will be used.

Let x denote the number of members of the government party and y the number of members of the neutral party. It follows from the conditions of the task that x and y are integers for which the following applies:

$x + y > \frac{2}{3}.253 = 168 + \frac{2}{3}$
$\Leftrightarrow x + y \geq 169, x + \frac{28}{100}y = 127$
$\Leftrightarrow 25x + 7y = 3175.$

Since $d(25,7)$, the equation has a solution. From the fact that

$7y = 3175 - 25x$

it follows that there exists $u \in Z$ such that

$3175 - 25x + (28x - 7.453) = 7u \leftrightarrow 3x = 7u - 4.$

So there exists m$\in Z$ such that

$3m = (7u - 4) + (-6 + 3) = u - 1$

$\Rightarrow u = 1 + 3m$
$\Rightarrow 3x = 7(1 + 3m) - 4 = 3 + 3.7m$
$\Rightarrow x = 7m$
$\Rightarrow 7y = 3175 - 25(1 + 7m) = 3150 - 7.25m$
$\Rightarrow y = 450 - 25m.$

So all integer solutions of the given equation are:

$x = 1 + 7m, y = 450 - 25m, m \in Z.$

From the conditions of the task, we get:

$x > 0$
$\Leftrightarrow 1 + 7m > 0$
$\Leftrightarrow m > -\frac{1}{7}; y > 0$
$\Leftrightarrow 450 - 25m > 0$
$\Leftrightarrow m < 18. x + y = 451 - 18m \geq 169$
$\Leftrightarrow 18m \leq 282$
$\Leftrightarrow m \leq 15 + \frac{2}{3}.$

For $m - 15$ we get:

$x - 106, y = 75.$

It follows that in the parliament, the government party has 106 deputies, neutrals 75 deputies and the opposition 72 deputies. By testing, we can easily convince ourselves that this solution meets the task's given conditions.

1.3. Extended Euclidean Algorithm

The reader has probably already encountered Euclid's algorithm for computing the largest common divisor of two positive integers. We will now present an extended version of this algorithm, which will serve as a tool for us many times in this text, for example, for proving one of the theorems or for solving linear Diophantine equations with two unknowns.

Given two positive integers $a, b, a > b$, the algorithm calculates their largest Common divisor $\delta = gcd(a, b)$ and simultaneously finds two integers α, β for which it holds

$$a\alpha + b\beta - \delta = 0. \tag{1.34}$$

Let's look at the individual steps of this algorithm:

E1: [Initialization.] Assign $\alpha' \leftarrow \beta \leftarrow 1, \alpha \leftarrow \beta' \leftarrow 0, \gamma \leftarrow a, \delta \leftarrow b.$

E2: [Division.] Let q and r respectively be the incomplete quotient and the remainder after dividing γ by δ.
(This means that $\gamma = q\delta + r, 0 \leq r < \delta.$)

E3: [Is the remainder zero?] If $r = 0$, the algorithm terminates, in which case it is

$$a\alpha + b\beta - \delta = 0.$$

E4: [New cycle.] Assign $\gamma \leftarrow \delta, \delta \leftarrow r, t \leftarrow \alpha', \alpha' \leftarrow \alpha, \alpha \leftarrow t - q\alpha,$ $t \leftarrow \beta', \beta' \leftarrow \beta, \beta \leftarrow t - q\beta$ and go back to step E2.

See e.g., Knuth [3] or Medvid [4].

Definition 1.3. Equality (1.34) is called Bezout's equality, and we call the numbers α, β Bezout coefficients.

Example 5. Find the greatest common divisor d of the numbers $1729, 551$ and find the numbers α, β so that

$$1769\alpha + 551\beta - \delta = 0.$$

We record the progress of the algorithm in the following table.

α'	α	β'	β	γ	δ	q	r
1	0	0	1	1769	551	3	116
0	1	1	−3	551	116	4	87
1	−4	−3	13	116	87	1	29
−4	5	2	13	87	29	3	0

So $\delta = 29, \alpha = 5, \beta = -16$ and indeed it is true that $1769.5 + 551.(-16) - 29 = 0$.

We will immediately use this algorithm in the proof of the following lemma, on which we will refer many times in the text.

Lemma 1.1. Let a, b, c be positive integers. If $gcd(a, b) = 1$ and at the same time $a \mid bc$, then $a \mid c$.

Proof. Since $gcd(a, b) = 1$, the extended Euclidean algorithm will find integers for the numbers a, b numbers α, β and are valid

$$a\alpha + b\beta - 1 = 0$$
$$ac\alpha + bc\beta - c = 0.$$

According to the assumptions $a \mid bc$, this means by Definition 2.1 that there exists a positive integer k and $bc = ka$ holds.

$$ac\alpha + ak\beta - c = 0$$
$$a(c\alpha + k\beta) = c$$

From the last equality, we see that $a \mid c$.

Let us first look at the question of the solvability of equation (1.1). Intuitively, we guess that it will have an integer solution only under some conditions. The following talks about her sentence.

Definition 1.4. We will say that equation (1.1) is in the basic form if and only when $gcd(a,b) = 1$.

Example 6. $2x + 4y + 7 = 0$

Since gcd(2, 4) = 2 and $2 \nmid 7$, this equation has no integer solutions.

Example 7. $3x + 8y - 5 = 0$

The equation has an integer solution, since $gcd(3,8) = 1$ and it holds that $1 \mid -5$.

Sometimes the following sentence that flows can also be useful for practical use from Theorem 1.1.

Theorem 1.2. Let equation (1.1) be given. Let there be a positive integer e such that $e \mid a, e \mid b$. If $e \nmid c$, then equation (1.1) does not have an integer solution.

Proof.

We can prove the variation of the theorem: If equation (1.1) has an integer solution, then $e \mid c$. That is, however (I) from the proof of Theorem 1. 1, where we now do not consider the greatest common divisor, but only the common divisor of the numbers a, b.

Example 8. 2586x + 4210y + 2361 = 0

According to Theorem 1.2, this equation does not have a solution in whole numbers, because the coefficients of the unknowns are even numbers, so they have a common divisor of 2, but the number 2361 is odd and thus is not divisible by 2.

Theorem 1.3. Let the equation (1.1) be given in the basic form. Let the integers x_0, y_0 are some solutions to this equation. Then the integers x, y are also solutions of this equation just when

$$x = x_0 - bt \; y = y_0 + at, \tag{1.35}$$

where t is an integer.

Proof.

The theorem is again stated in the form of equivalence, so we must prove the following:

(I) If $x = x_0 - bt$ and $y = y_0 + at$, where x_0, y_0 is an integer solution of equation (1.1), t is an integer, then x, y is also a solution.

(II) If x, y is an integer solution of equation (1.1), then

$x = x_0 - bt \; y = y_0 + at$, where t is an integer.

ad(I) We want to verify that numbers of the form x, y are solutions of equation (1.1), let's substitute and calculate.

$$\begin{aligned} ax + by + c &= a(x_0 - bt) + b(y_0 + at) + c \\ &= ax_0 - abt + by_0 + abt + c \\ &= ax_0 + by_0 + c = 0 \end{aligned}$$

The last equality follows from the fact that x_0, y_0 is a solution.

ad(II) We know that x_0, y_0 and x, y are solutions of equation (1.1), so the equalities hold

$$ax + by + c = 0 \; ax_0 + by_0 + c = 0.$$

Let's subtract the second equality from the first and calculate.

$$a(x - x_0) + b(y - y_0) = 0$$

$$b(y - y_0) = a(x_0 - x) \tag{1.36}$$

Since we consider equation (1.1) in the basic form, i.e., $gcd(a, b) = 1$, it must by Lemma 1.1 b | $(x_0 - x)$. Again, this means that it exists integer t such that $x_0 - x = bt$. From there, we have

$$x = x_0 - bt.$$

By substituting bt for $x_0 - x$ in equation (1.36) we obtain

$$y - y_0 = \frac{a}{b}bt$$
$$y - y_0 = at$$
$$y = y_0 + at.$$

The theorem says that we actually only need to find one arbitrary solution to equation (1.1) in the basic form and we immediately know all other solutions of this equation and are of the form (1.35).

Now we will separately solve the case a $=$ b. If both coefficients are the same, we can equation (1.1), provided that it has a solution, convert it into the form

$$x + y + e = 0,$$

where e is an integer for which $c = a \cdot e$. Solving this equation is very easy, just choose one unknown, e.g., y, as an integer parameter, $y = t$, where t is an integer, and add the second unknown. In total

$$x = -e - t\ y = t.$$

It is enough for us to solve the case when $a > b$. In particular, it can also happen that $b = 1$, but then we solve the equation

$$x + y + c = 0.$$

Solving this equation is again simple;

$$x = t\ y = -c - at,$$

where t is an integer.

So, we will deal with equations where $a > b > 1$.

1.4. Solution Method by the Extended Euclidean Algorithm

We will use the already known extended Euclidean algorithm. For an overview, I will remind you of our task.

We want to solve the equation

$$a\,x + by + c = 0, \tag{1.37}$$

where a, b, c are integers, $a > b > 1$.

Let's apply the extended Euclidean algorithm to the coefficients a and b. The result will be Bezout's equality

$$a\alpha + b\beta - \delta = 0, \tag{1.38}$$

where α, β are integers, $\delta = gcd(a, b)$. It now decides whether the equation has an integer solution based on whether $d \mid c$. If this is not the case, he does not have, according to theorem 1.1 solving equations in whole numbers. If, on the contrary, $\delta \mid c$, there exists throughout number k such that $c = k\delta$. We will want to compare the obtained equality (1.38) with equation (1.37), we must therefore multiply it by the number $-k$ and we get

$$a(-k\alpha) + b(-k\beta) + k\delta = 0$$

$$a(-k\alpha) + b(-k\beta) + c = 0.$$

From here, comparing with equation (1.37), we see one integer solution

$$x_0 = -k\alpha\, y_0 = -k\beta.$$

We will use Theorem 1.3 to describe all solutions. However, its assumption is the inequality of the coefficients standing with the unknowns. We must therefore divide them by their largest common divisor δ and we obtain all solutions of equation (1.37) of the form

$$x = -k\alpha - \frac{b}{\delta} t \; y = -k\beta + \frac{a}{\delta} t,$$

where t is an integer.

If equation (3.5) is in the basic form, i.e., $\delta = 1$, then by solving this equation, we can shape numbers

$$x = -c\alpha - bt \; y = -c\beta + at,$$

where t is an integer.

Let's show this procedure with concrete examples.

Example 9. $25x + 17y + 4 = 0$

According to the instructions, let's apply the extended Euclid algorithm to the numbers $25, 17$. Progress, we will always record the algorithm in a table.

α'	α	β'	β	γ	δ	q	r
1	0	0	1	25	17	1	8
0	1	1	−1	17	8	2	1
1	−2	−1	3	8	1	8	0

We calculated that $\alpha = -2$ and $\beta = 3$; we thus obtain the following Bezout equality

$$25(-2) + 17 \cdot 3 - 1 = 0,$$

which just needs to be multiplied by the number -4.

$$25\,8 + 17(-12) + 4 = 0$$

We found one integer solution

$$x_0 = 8 \quad y_0 = -12,$$

which, according to Theorem 3.3, is enough for us to describe all other solutions, namely

$$x = 8 - 17t \; y = -12 + 25t,$$

where t is an integer.

Example 10. $1025x - 241y - 26 = 0$

Since the coefficient of the unknown y is negative, we must introduce the substitution

$$y = -z$$

and thus we get the equation

$$1025x + 241z - 26 = 0,$$

where we already have both positive coefficients, and we can apply the extended Euclidean algorithm.

α'	α	β'	β	γ	δ	q	r
1	0	0	1	1025	241	4	61
0	1	1	−4	241	61	3	58
1	−3	−4	13	61	58	1	3
−3	4	13	−17	58	3	19	1
4	−79	−17	336	3	1	3	0

We thus obtain Bezout's equality

$$1025(-79) + 241 \cdot 336 - 1 = 0,$$

which we just need to multiply by the number 26.

$$1025(-2054) + 241 \cdot 8736 - 26 = 0$$

We see that

$$x_0 = -2054\ z_0 = 8736,$$

and according to theorem 1.3

$$x = -2054 - 241t\ z = 8736 + 1025t.$$

But we have to go back to the substitution $y = -z$ to find a solution to the original equation.

In total

$$x = -2054 - 241t\ y = -8736 - 1025t,$$

where t is an integer.

Example 11. $136x + 85y - 187 = 0$

Again, we perform the extended Euclidean algorithm for the coefficients $136, 85$.

α'	α	β'	β	γ	δ	q	r
1	0	0	1	136	85	1	51
0	1	1	−1	85	51	1	34
1	−1	−1	2	51	34	1	17
−1	2	2	−3	34	17	2	1

We get Bezout's equality

$$136(2) + 85(-3) - 17 = 0 \tag{1.39}$$

and we see that $gcd(136, 85) = 17$. We need to see if the equation has an integer solution; it must be true that 17 | 187, which is true because

$$187 = 11 \cdot 17.$$

We must multiply the equation (1.39) by number 11.

$$136(22) + 85(-33) - 187 = 0$$

One solution is the numbers $x_0 = 22, y_0 = -33$. To describe all solutions, we have to use Theorem 3.3, but the assumption of this theorem is that $gcd(a, b) = 1$, which we do not have in our example. We must therefore divide the numbers $136, 85$ by their greatest common divisor, i.e., 17 and we get the numbers $8, 5$. So all solutions are of the form

$$x = 22 - 5t\ y = -33 + 8t,$$

where t is an integer.

Exercises. Use the extended Euclidean algorithm to solve the following equations:

$$27x - 34y - 5 = 0$$

$$-51x - 221y - 85 = 0$$

$$2432x + 237y + 3 = 0$$

$$65x + 156y + 11 = 0$$

1.5. The Method of Solving by Congruence

Unfortunately, we cannot do without the introduction of the necessary apparatus, which we need to pay more attention to, because we will use it later when solving linear Diophantine equations with n unknowns, but also for equations that are linear with respect to at least one unknown. First, we will introduce the concept of congruence, and then we will show how to count by congruences. See e.g., Verdal [8], Chionglo [1] or Jones [2].

Definition 1.5. If two integers a, b have m after division by a positive integer, the same remainder r, where $0 \leq r < m$, we say that a is congruent to b modulo m and we write symbolic

$$a \equiv b \ (mod\ m).$$

Otherwise, we say that a is not congruent to b modulo m and write

$$a \not\equiv b \ (mod\ m).$$

We call the number a the left side of the congruence, the number b the right side of the congruence.

Theorem 1.4. Let a, b be integers and m be a positive integer. Then the following conditions are equivalent:

$(I)\ a \equiv b\ (mod\ m)$.

$(II)\ m|\ (a - b)$.

$(III)\ a = b + mk$,

where k is an integer.

Proof.

We do it by proving the following three statements:

(i) If (I), then (II).

(ii) If (II), then (III).

(iii) If(III), then(I).

ad(i) $a \equiv b\ (mod\ m)$ by Definition 1.5 means

$a = q_1m + r$

$b = q_2m + r,$

where q_1, q_2 are integers. Subtracting the second equation from the first gives

$$a - b = (q_1 - q_2)m,$$

which, of course, means that $m \mid (a - b)$.

ad (ii) Since $m \mid (a - b)$, there must exist an integer k such that

$a - b = km,$

by modification, we get

$a = b + mk.$

ad(iii) If we divide b with the remainder by m, we get

$$b = qm + r,$$

where q, r are integers and $0 \leq r < m$.

Substituting into the assumption $a = b + mk$ we obtain

$$a = (q + k)m + r,$$

i.e., the number a has the same remainder after division by the number m as the number b, which we had to prove.

Theorem 1.5. Let a, b, c be integers. Then for every positive integer m:

$(I)\ a \equiv a\ (mod\ m).$

$(II)\ If\ a \equiv b\ (mod\ m), then\ b \equiv a\ (mod\ m).$

$(III)\ If\ a \equiv b\ (mod\ m) and\ b \equiv c\ (mod\ m), then\ so\ is$
$a \equiv c\ (mod\ m).$

Proof.

ad(I) For every positive integer $m \mid 0$, which we can write $m \mid (a - a)$, that by Theorem 1.4 is $a \equiv a\ (mod\ m)$.

ad(II) By assumption $a \equiv b\ (mod\ m)$, must $m \mid (a - b)$, that is, there exists an integer k such that $a - b = km$. After multiplying this equality by -1 we get

$$b - a = -km,$$

that is,

$$m \mid (b - a),$$

so

$$b \equiv a \pmod{m}.$$

ad(III) Under the assumptions $a \equiv b \pmod{m}, b \equiv c \pmod{m}$, there exist integers k, l such that it applies

$$a - b = km$$
$$b - c = lm.$$

After adding these equations, we get

$$a - c = (k + l)m,$$

i.e.,

$$a \equiv c \pmod{m}.$$

The following theorems talk about how to calculate congruences.

Theorem 1.6. Let a, b, c, m be integers, $m > 0$. If $a \equiv b \pmod{m}$, then is also

(I) $a + c \equiv b + c \pmod{m}$,

(II) $a - c \equiv b - c \pmod{m}$,

(III) $ac \equiv bc \pmod{m}$.

Proof.

ad(I) By assumption $m \mid (a - b)$. This is the same as $m \mid (a + c - c - b)$. After adjustment, we have $m \mid [(a + c) - (b + c)]$, which of course, means that

$$a + c \equiv b + c \pmod{m}.$$

Ad (II) According to the assumption $m \mid (a - b)$. This is the same as

$$m \mid (a - c + c - b).$$

After adjustment, we have $m \mid [(a - c) - (b - c)]$, which of course, means that

$$a - c \equiv b - c \ (mod\ m).$$

Ad (III) Since $m \mid (a - b)$, holds $m \mid (a - b)c$, so $m \mid (ac - bc)$, hence

$$ac \equiv bc \ (mod\ m).$$

If we have a given congruence with a given module m, we can add to it, or subtract from it any integer or multiply both sides of a congruence by any integer, as with an equality.

Theorem 1.7. Let a, b, c, d be integers. Let m be a positive integer. If it is $a \equiv b \ (mod\ m)$ and at the same time $c \equiv d \ (mod\ m)$ is then also

$(I)\ a + c \equiv b + d \ (mod\ m)$

$(II)\ a - c \equiv b - d \ (mod\ m$ $(III)\ ac \equiv bd \ (mod\ m)$.

Proof.

ad(I) From the congruences

$a \equiv b \ (mod\ m),$
$c \equiv d \ (mod\ m)$

we obtain according to (I) in Theorem 1.6

$a + c \equiv b + c \ (mod\ m)$
;+;; $c + b \equiv d + b \ (mod\ m)$.

Now just use (III) from Theorem 1.6 and we get

$$a + c \equiv b + d \ (mod\ m).$$

ad(II) From the congruences

$$a \equiv b \ (mod\ m),$$
$$c \equiv d \ (mod\ m)$$

we obtain according to (II) in Theorem 1.6

$$a - c \equiv b - c \ (mod\ m)$$
$$c - b \equiv d - b \ (mod\ m).$$

We multiply the congruence on the right by the number -1 according to (III) from Theorem 1.6. That's how we get it congruence

$$b - c \equiv b - d \ (mod\ m)$$

and now it is enough to use (III) from Theorem 1.6 again.

ad(III) From the congruences

$$a \equiv b \ (mod\ m),$$
$$c \equiv d \ (mod\ m)$$

we obtain according to (III) in Theorem 1.6

$$ac \equiv bc \ (mod\ m)$$
$$cb \equiv db \ (mod\ m).$$

Now it is enough to use (III) from Theorem 1.5 and we get

$$ac \equiv bd \ (mod\ m).$$

Thus, congruences according to the same module can be added together, subtracted and multiplied. A consequence of this theorem is that we can transfer any addendum from one side congruence squared, but with the opposite sign, and we can add a different and arbitrary multiple of the modulus to each side of the congruence.

Theorem 1.8. If $a \equiv b \ (mod\ m)$, then for every positive integer k holds

$$a^k \equiv b^k \ (mod\ m).$$

Proof.

We can prove the theorem by mathematical induction. We need to prove the following:

(I) We prove that the theorem holds for $k = 1$.

(II) We prove that $a^{k+1} \equiv b^{k+1} \ (mod\ m)$holds, provided

$$a^k \equiv b^k \ (mod\ m).$$

ad(I) For $k = 1$, we get $a^1 \equiv b^1 \ (mod\ m)$, which is, of course, an assumption of the theorem.

ad(II) Assume that the theorem holds for k, i.e.,

$$a^k \equiv b^k \ (mod\ m).$$

If we put $c = a^k$ and d=b^k in Theorem 3.9, we get according to point (III) of this theorem

$$a. a^k \equiv b. b^k \ (mod\ m)$$

which is

$$a^{k+1} \equiv b^{k+1} \ (mod\ m).$$

Thus, we can exponentiate both sides of the congruence by a positive integer, similarly to equality.

Theorem 1.9. Let a, b, c, m be integers, $m > 0$. Let $gcd(c, m) = 1$. Let finally,

$ac \equiv bc \ (mod\ m).$

Then also

$a \equiv b \ (mod\ m).$

Proof.

Since

$$ac \equiv bc \ (mod\ m),$$

$m \mid (ac - bc),$

i.e.,

$m \mid (a - b)c$

and since

$gcd(c, m) = 1,$

must

$m \mid (a - b),$

which means

$a \equiv b \ (mod\ m).$

Therefore, we can abbreviate the congruence by a number that is coprime with the given modulus congruence.

Theorem 1.10. Let a, b, c, m be integers, $c > 0, m > 0$. Congruence $a \equiv b$ (mod m) is equivalent to the congruence $ac \equiv cb \ (mod\ mc)$.

Proof.

We need to prove two things:

$(I)\ a \equiv b \ (mod\ m) \Rightarrow ac \equiv cb \ (mod\ mc),$
$(II)\ ac \equiv cb \ (mod\ mc) \Rightarrow a \equiv b \ (mod\ m).$

ad(I) Since $a \equiv b \ (mod\ m)$, there exists an integer k such that

$$a - b = km,$$

we can multiply this equality by a positive integer c,

$$ac - bc = k(mc),$$

hence

$$ac \equiv cb \ (mod\ mc).$$

ad(II) Since

$$ac \equiv cb \ (mod\ mc),$$

there exists an integer k such that

$$ac - bc = kmc,$$

we can divide this equality by a positive integer c,

$$a - b = km,$$

hence

$$a \equiv b \ (\text{mod m}).$$

We will show how congruences help us in solving linear Diophantine equations with two unknowns. We will return to the solution of equation (1.37) again, as long as we feel like it and consider this equation in the following form

$$ax + by = c, \tag{1.40}$$

Moreover, we no longer require $a > b > 1$, the numbers a, b are arbitrary, but non-zero integers, and different from 1. Any solution x, y after substituting into equation (1.40) must necessarily satisfy congruence according to any modulus m, where m is a positive integer, i.e.,

$$ax + by \equiv c \ (mod\ m).$$

Let's take the module as the absolute value of one of the coefficients, e.g., let $m = |b|$, we will receive

$$ax + by \equiv c \ (mod\ |b|),$$

because we can attribute, or subtract multiples of the modulus from the congruence, we get

$$ax \equiv c \ (mod\ |b|). \tag{1.41}$$

We obtained a linear congruence with one unknown.

Theorem 1.11. Let m be a positive integer, u, v be integers. Let $gcd(u, m) = \delta$.

If $\delta \mid v$, then congruence

$$ux \equiv v \ (mod\ m) \tag{1.42}$$

has a solution.

Proof.

Because $\delta \mid v$, is $v = k \cdot \delta$, where k is an integer. Using the extended Euclidean algorithm, we can find integers α, β such that:

$$\alpha u + \beta m = \delta$$

$$(\alpha k)u + (\beta k)m = k\delta$$

$$(\alpha k)u + (\beta k)m = v$$

$(\beta k)m = v - (\alpha k)u$

From there $m \mid (v - (\alpha k)u)$, which is

$$(\alpha k)u \equiv v \,(mod\, m). \quad (1.43)$$

Comparing with equation (1.42), we see the solution $x = \alpha k$.

Theorem 1.12. Let m be a positive integer, u, v be integers. Let $gcd(u, m) = \delta$.

Congruence

$$ux \equiv v \,(mod\, m) \quad (1.44)$$

is solvable only if $d \mid v$. In this case, x is a solution of the congruence (1.44), precisely when

$$x \equiv v_1 \cdot u_1^{\varphi(m_1)-1} \,(mod\, m_1), \quad (1.45)$$

where $u_1 = ud, v_1 = \frac{v}{\delta}, m_1 = \frac{m}{\delta}$ and ϕ is the Euler function.

Compared to Theorem 1.11, this Theorem shows us how the congruence (1.44) can be converted to (1.45), from which, according to Theorem (1.4), one can see the solution to the congruence (1.44) of the form

$$x = v_1 \cdot u_1^{\varphi(m_1)-1} + tm_1$$

where t is an integer. However, we are working here with the Euler function, for which there is no efficient algorithm for its calculation if we do not know the prime factorization of its argument, and in addition, the number $u_1^{\varphi(m_1)-1}$ is usually very high. For these reasons, we will suffice with Theorem 1.11, which guarantees us the existence of a solution to the congruence (1.42), which we will learn to convert to the form $x \equiv q \,(mod\, m)$, where q is an integer, using the allowed adjustments.

Example 12. $5x \equiv 36 \,(mod\, 7)$

First, we write out the entire calculation and then comment on the individual steps. This according to Theorem 1.11, the congruence has a solution, since $gcd(5,7) = 1$ and $1 \mid 36$.

$$5x \equiv 36 \ (mod\ 7)$$

$$5x \equiv 15 \ (mod\ 7)$$

$$x \equiv 3 \ (mod\ 7)$$

$$x = 3 + 7t,$$

where t is an integer.

During the calculation, we gradually used the subtraction of three times the modulus from the left side of the congruence (corollary of Theorem 1.7), dividing by five, which is possible because $5 \nmid 7$ (Theorem 1.9) and finally the equivalent notation (Theorem 1.4).

In the following two examples, we will concretely demonstrate the use of congruences in the solution equation (1.40).

Example 13. $14x + 23y = 17$

Let x, y be solutions, then they satisfy congruence

$$x + 23y \equiv 17 \ (mod\ 14),$$

it is often advantageous to choose the smallest possible modulus, i.e., the smallest of the coefficients. By editing, we gradually obtain:

$$23y \equiv 17 \ (mod\ 14)$$

$$9y \equiv 3 \ (mod\ 14)$$

$$3y \equiv 1 \ (mod\ 14)$$

$$3y \equiv 15 \ (mod\ 14)$$

$$y \equiv 5 \ (mod\ 14)$$

In the modifications, we gradually used the subtraction of one times the modulus, division by the number 3, which is possible because the number 3 is not commensurate with the module, adding one multiple of the module to the left side of the congruence, and another division by the number 3.

According to Theorem 1.4, the last congruence gives us $y = 5 + 14t$, where t is an integer.

By substituting into the entered equation, we calculate the unknown x.

$$14x + 23(5 + 14t) = 1$$

$$14x = -98 - 23 \cdot 14t$$

$$x = -7 - 23t$$

In total, then

$$x = -7 - 23t\ y = 5 + 14t,$$

where t is an integer.

Example 14. $18x - 25y = -8$

If the numbers x, y are a solution, then congruence must satisfy, for example

$$18x - 25y \equiv -8\ (mod\ 18)$$

and by successive adjustments, we get:

$$-25y \equiv -8\ (mod\ 18)$$

$$25y \equiv 8\ (mod\ 18)$$

$$25y \equiv -10\ (mod\ 18)$$

$$5y \equiv -2\ (mod\ 18)$$

$$5y \equiv -20\ (mod\ 18)$$

$$y \equiv -4 \ (mod\ 18)$$

Hence $y = -4 + 18t$, where t is an integer. By substituting into the entry, we get

$$18x - 25 \cdot (-4 + 18t) = -8.$$

In total, then

$$x = -6 + 25t \ y = -4 + 18t,$$

where t is an integer.

These two examples show how easy and almost effortless the solution is; that's all, you only need to grasp the concept of congruence and be able to solve linear congruences about one unknown.

A complication arises if the given Diophantine equation has large coefficients; then the congruence modulus is also large, and finding the solution becomes more difficult.

The following theorems will help us with the case where the modulus is not prime, i.e., it will go right as the product of several positive integers. It turns out that we can then convert this congruence into a system of multiple congruences according to the modules that are the factors of the original module.

We must therefore prove the equivalence of the original congruence and the resulting system of congruences, and then provide instructions on how to solve a system of linear congruences with one unknown.

First, two auxiliary sentences.

Lemma 1.2. Let x, y, z be integers. If $x \mid y$ and at the same time $y \mid z$, then $x \mid z$.

Proof.

We want to prove that $x \mid z$, which means we need to find an integer t such that it holds $z = t \cdot x$. By assumption, $x \mid y$ and $y \mid z$, so there must be integers k, l such that that $y = k \cdot x$ and $z = l \cdot y$. By substituting into the last relation for y, we get $z = l \cdot k \cdot x$, from there by choosing $t = k \cdot l$ we found the sought t.

Definition 1.6. Let a, b be positive integers. We will say that the positive integer γ is the least common multiple of numbers a, b if and only if:

(*I*) $a \mid \gamma \wedge b \mid \gamma$,

(*II*) for all positive integers e: $(a \mid e \wedge b \mid e) \Rightarrow \gamma \mid e$.

We will write down the fact that γ is the least common multiple of the numbers a, b

$\gamma = lcm(a, b)$.

If only (*I*) holds, we call the number γ a common multiple of the numbers a, b.

Theorem 1.13. Let u, v be integers. Let k be a positive integer. Let m_i have them be positive integers for all $i = 1, 2, \ldots, k$. Finally, let $m = lcm(m_1, m_2, \ldots, m_k)$.

Then the congruence $u \equiv v \ (mod\ m)$ is satisfied if and only if the congruences are satisfied $u \equiv v \ (mod\ m_i)$ for all $i = 1, 2, \ldots, k$.

Proof.

Since the sentence is stated in the equivalence form, we need to prove two things:

(*I*) If $u \equiv v \ (mod\ m)$, then $u \equiv v \ (mod\ m_i)$, for all $i = 1, 2, \ldots, k$.

(*II*) If for all $i = 1, 2, \ldots, k$ holds for $u \equiv v \ (modm_i)$, then $u \equiv v \ (mod\ m)$.

ad(*I*) If $u \equiv v \ (mod\ m)$, then must $m \mid (u - v)$. Let $i = 1, 2, \ldots, k$. Because m is the least common multiple of the numbers m_i, then $m_i \mid m$ and as a consequence of transitivity, the relation divides, i.e., of Lemma 3.2, must also $m_i \mid (u - v)$, which of course, means that

$u \equiv v \ (mod\ m_i)$.

ad(*II*) If for all $i = 1, 2, \ldots, k$ is $u \equiv v \ (mod\ m_i)$, must $m_i \mid (u - v)$.

So, the number $u - v$ must be a common multiple of the numbers $m_i, i = 1, 2, \dots, k$, and hence must be divisible by their least common multiple, i.e., the number m. Because $m| \ (u - v)$, must be $u \equiv v \ (mod \ m)$.

For example, the congruence $7x = 2 \ (mod \ 15)$ is, according to this theorem, equivalent to the set of congruences

$$7x \equiv 2 \ (mod \ 5)$$

$$7x \equiv 2 \ (mod \ 3).$$

So, there was a decrease in modulus, but an increase in equations. We must learn to solve these broken systems.

Consider the solvable congruence $ax \equiv b \ (mod \ n)$ and perform a prime factorization of numbers n:

$$n = p_1^{\alpha_1} p_2^{\alpha_2} \dots p_k^{\alpha_k}$$

where p_i are mutually different prime numbers, α_i are positive integers, for all $i = 1, 2, \dots, k$. Because $lcm(p_1^{\alpha_1} p_2^{\alpha_2} \dots p_k^{\alpha_k}) = n$, this congruence is according to Theorem 3.15 equivalent to the system of congruences

$$ax \equiv b \ (mod \ p_1^{\alpha_1})$$

$$ax \equiv b \ (mod p_2^{\alpha_2})$$

.

.

$$ax \equiv b \ (mod \ p_k^{\alpha_k}).$$

Since the original congruence is solvable and is equivalent to the resulting system, each of these congruences is solvable. Then, of course, we can use the enabled system adjustments to convert to shape

$$x \equiv c_1 \ (mod \ p_1^{\alpha_1})$$

$$x \equiv c_2 \ (mod p_2^{\alpha_2})$$

.

.

$$x \equiv c_k \ (mod\ p_k^{\alpha_k}).$$

where c_i are suitable integers, $i = 1, 2, \ldots, k$.

Let us first consider $k = 2$, so we are looking for a solution to the following system

$$x \equiv c_1 \ (mod\ m_1)$$

$$x \equiv c_2 (mod m_2). \tag{1.46}$$

The following theorem, known as the Chinese Remainder Theorem, tells about the solution of the system of congruences (1.46) looks like.

Theorem 1.14 (Chinese Remainder Theorem). Let m be a positive integer, moreover, such that $m = m_1 \cdot m_2$, where m_1, m_2 are positive integers,$m_1 \geq 2, m_2 \geq 2$, $gcd(m_1, m_2) = 1$. See e.g., Weisstein [9].

System of linear congruences

$$x \equiv c_1 \ (mod\ m_1)$$
$$x \equiv c_2 (mod\ m_2)$$

always has a solution, and this solution is uniquely determined modulo m.

Proof.

The solution to the first congruence is the form numbers

$$x = c_1 + m_1 t, \tag{1.47}$$

where t is an integer. Substituting this solution into the second congruence, we have

$$c_1 + m_1 t \equiv c_2 \ (mod\ m_2)$$

$$m_1 t \equiv c_2 - c_1 \ (mod\ m_2)$$

We obtained a linear congruence for t, which according to Theorem 1.11 has a solution, because by assumption $gcd(m_1, m_2) = 1$ and $1 \mid (c_2 - c_1)$.

Therefore, we can adjust this congruence to the form $t \equiv c_3 \ (mod\, m_2)$, where c_3 is some integer, hence

$$t = c_3 + s\, m_2 .$$

Substitution to (1.47)

$$x = c_1 + m_1 \cdot (c_3 + s\, m_2)$$

$$x = c_1 + m_1 c_3 + s\, m_1\, m_2$$

$$x = c + s\, m_1\, m_2$$

$$x \equiv c \ (mod\, m_1\, m_2)$$

$$x \equiv c \ (mod\, m), \qquad (1.48)$$

where we denoted $c = c_1 + m_1 c_3$.

Note that the obtained congruence, which is a solution of the system (1.46) is of the same form as the original congruence. The proof of the Chinese Remainder Theorem is also a guide to how to solve a system of linear congruences about one unknown. So, if we have a system of more than two congruences, then we replace the first two with a single congruence according to Theorem 3.16 and for it we take the third congruence, etc. If we solve the system k congruence, so after $k - 1$ steps we find the solution of the entire system.

We will show everything in the following example.

Example 15. $238x - 171y = 23$

We can see that the coefficients are relatively large numbers, which would lead to inconvenient calculations with large moduli. Therefore, we will use Theorem 1.13. The solution must meet

$$238x - 171y \equiv 23 \ (mod\ 171)$$

$$67x \equiv 23 \ (mod\ 171).$$

Since $171 = 9 \cdot 19$, we can replace the last congruence with the system

$$67x \equiv 23 \ (mod\ 9)$$

$$67x \equiv 23 \ (mod\ 19).$$

We find a solution to the first of them

$$67x \equiv 23 \ (mod\ 9)$$

$$4x \equiv -4 \ (mod\ 9)$$

$$x \equiv -1 \ (mod\ 9)$$

$$x = -1 + 9s,$$

where s is an integer. We insert this solution into the second congruence.

$$67(-1 + 9s) \equiv 23 \ (mod\ 19)$$

$$-67 + 603s \equiv 23 \ (mod\ 19)$$

$$603s \equiv 90 \ (mod\ 19)$$

$$14s \equiv 14 \ (mod\ 19)$$

$$s \equiv 1 \ (mod\ 19)$$

$$s = 1 + 19t,$$

where t is an integer. Substituting for s, we have $x = 8 + 171t$. Now it remains to calculate y, that again, by substituting the solution for x into the input, we get

$$y = 11 + 238t.$$

Exercises. Solve the following Diophantine equations using congruences.

$(a)\ 41x + 12y = 5$

$(b)\ 9x + 4y = 12$

$(c)\ 72x - 27y = 18$

$(d)\ 826x - 1074y = 2808$

References

[1] Chionglo, J. F. How I solved the linear congruence 25x = 15 (mod 29). doi:10.13140/RCR2.2.1476856324 (2016).

[2] Jones, G. A., & Jones, J. J. *Congruences*. In J. J. Jones & G. A. Jones, *Elementary Number Theory* (Springer Undergraduate Mathematics Series). London: Springer. (1991).

[3] Niven, I., Zuckerman, H. S., & Montgomery, H. L. *An Introduction to the Theory of Numbers*. New York: Wiley. (1998).

[4] Knuth, D. E. *The Art of Computer Programming*, Vol. 2, Chapter 4. Addison-Wesley. (1981).

[5] Medvid', V. Algorithms of solving the Diophantine equation. In *ICERI 2023: 16th International Conference of Education, Research and Innovation, 13–15 November 2023, Seville, Spain* (conference proceedings). (2023).

[6] Mordell, L. J. *Diophantine Equations*. Pure and Applied Mathematics, Vol. 30. Academic Press. ISBN 0-12-506250-8. (1969).

[7] Shallit, J. Origins of the analysis of the Euclidean algorithm. *Historia Mathematica, 21*(4), 401–419. doi:10.1006/hmat.1994.1031(1994).

[8] Schmidt, W. M. *Diophantine Approximations and Diophantine Equations*. Lecture Notes in Mathematics, Vol. 1467. Berlin: Springer-Verlag. ISBN 3-540-54058-X. (1991).

[9] Verdal, E. Solutions of some classes of congruences. *The Teaching of Mathematics, 9*(1), 41–44. (2006).

[10] Weinberger, P. On Euclidean rings of algebraic integers. In *Proceedings of Symposia in Pure Mathematics, Vol. 24* (pp. 321–332). Providence, RI: American Mathematical Society. doi:10.1090/pspum/024/0337902 (1973).

[11] Weisstein, E. W. (1999). Chinese remainder theorem. From *MathWorld – A Wolfram Resource*. Retrieved from https://mathworld.wolfram.com/ChineseRemainderTheorem.html (1999).

Chapter 2

Diophantine Linear Equations with Three Unknowns

Abstract

In this chapter, the author presents in detail two novel methods for solving complex Diophantine equations with three unknowns, providing explicit solutions.

A linear Diophantine equation with three unknowns of the form

$$ax + y + z = d$$

can be solved using a method analogous to that for equations with two unknowns. The procedure is illustrated through the construction of an algorithm for solving such equations, along with its application to specific examples.

This approach to solving linear Diophantine equations with three unknowns can be easily extended to equations with any number of unknowns. The condition for solvability is also analogous: it depends on whether the greatest common divisor (GCD) of the coefficients of the unknowns divides the constant term.

Educational goal:

These methods have been tested, similarly to methods for two-variable Diophantine problems, in secondary school mathematics classes as well as at university-level mathematics faculties. Students mastered these methods relatively easily and were subsequently able to solve complex problems they had previously been unable to handle. Therefore, it is recommended to incorporate these methods into study materials, especially for talented students, to foster a deeper understanding of the topic.

Keywords: Diophantine equation of the second degree, Euclidean sequencing algorithm, Euclidean algorithm, algorithm, three unknowns, condition, unknown, coefficient, the greatest common divisor

2.1. Introduction

The linear Diophantine equation with three unknowns in the form

$$ax + by + cz = d\,, \tag{2.1}$$

where $a, b, c, d \in Z \; and \; a \neq 0, b \neq 0, c \neq 0, d \neq 0.$

Rather than directly seeking a solution to equation (2.1), we first determine the conditions under which it is solvable (see e.g., Medvid' [1] and Rania [2]).

Condition of solvability

Equation (2.1) can be rewritten as

$$ax + by = d - cz. \tag{2.2}$$

If we regard equation (2.2) as a Diophantine equation in the unknowns x and y, the necessary and sufficient condition for solvability is that the greatest common divisor of a and b divides the right-hand side, i.e.,

$$\delta = gcd(a, b) \mid (d - cz)$$

In other words, equation (2.1) is solvable if there exists an integer t such that

$$d - cz = \delta t.$$

This equation can be rewritten as

$$\delta t + cz = d. \tag{2.3}$$

Equation (2.3) is itself a Diophantine equation in the unknowns z and t. Therefore, equation (2.1) is solvable if and only if equation (2.3) is solvable.

The solvability condition for equation (2.3) is that the greatest common divisor

$$\varepsilon = gcd(c, \delta)$$

must divide d. From the theory of prime factorization of integers, it follows that if

$$\delta = gcd(a, b) \text{ and } \varepsilon = gcd(c, \delta)$$

then

$$\varepsilon = gcd(a, b, c).$$

Hence, ε is the greatest common divisor of the numbers $a, b,$.and c. From this, we obtain an important property:

The necessary and sufficient condition for the solvability of equation (2.1) is that the greatest common divisor of a, b, and c divides d:

$$\delta = gcd(a, b, c) \Rightarrow \delta \mid d.$$

2.2. Algorithm of the Solution of the Diophantine Equation with Three Unknowns

Assume that equation (2.1) satisfies the solvability condition:

$$\delta = gcd(a, b, c) \Rightarrow \delta \mid d.$$

We begin by selecting the unknown variable with the smallest coefficient. Without loss of generality, let this be a. Then equation (2.1) can be written as

$$ax = d - by - cz \tag{2.4}$$

It follows that $a \mid (d - b.y - c.z)$. Therefore, there exists an integer $u \in Z$ such that

$$au = -by - cz + d. \tag{2.5}$$

To simplify the expression, we strategically add or subtract appropriate multiples of a, aiming to reduce the absolute values of the coefficients $-b, -c$, and the constant term d. This yields

$$au = -by - cz + d + (k_1ay + k_2az + k_3a). \tag{2.6}$$

Rewriting the right-hand side, we obtain

$$au = (k_1a - b)y + (k_2a - c)z + (k_3a + d) \tag{2.7}$$

Next, we select the smaller of the absolute values of the coefficients $min\{|k_1a - b|, |k_2a - c|\}$. Without loss of generality, assume the smaller value is $k_1a - b$. We then isolate the term $(k_1a - b)y$ from equation (2.7):

$$(k_1a - b)y = au - (k_2a - c)z - (k_3a + d). \tag{2.8}$$

Clearly,

$$(k_1a - b) \mid (au - (k_2a - c)z - (k_3a + d)).$$

So there exists an integer $v \in Z$ such that

$$(k_1a - b)v = au - (k_2a - c)z - (k_3a + d). \tag{2.9}$$

To further reduce the coefficients, we add suitable integer multiples of $(k_1a - b)$ to the right-hand side of equation (2.9), aiming to minimize the magnitude of the resulting coefficients. That is,

$$(k_1a - b)v =$$
$$au + [l_2(k_1a - b) - (k_2a - c)]z + [l_3(k_1a - b) - (k_3a + d)].$$

This procedure is repeated iteratively until the coefficient of the unknown z is reduced to either 1 or -1, thereby simplifying the equation.

The resulting equation can now be expressed in the simplified form

$$(k_1a - b)v = au + z + f,$$

which can be rearranged to isolate z:

$$z = (k_1a - b)v - au - f. \tag{2.10}$$

Substituting the expression for z from equation (2.10) back into equation (2.8) allows us to solve explicitly for y.

Finally substituting both y and z into the right-hand side of equation (2.4) enables us to solve explicitly for the remaining unknown x.

We now proceed to illustrate this algorithm with a concrete example.

Example 1. Find all integer solutions to the equation:

$$18x - 12y + 15z = 6 \tag{2.11}$$

Solution. First, we check whether the equation admits integer solutions. We compute the greatest common divisor of the coefficients:

$$d(18,-12,15) = 3, 3|6.$$

Since 3 divides the right-hand side, the equation has integer solutions.. To simplify, we divide both sides by 3:

$$6x - 4y + 5z = 2.$$

We now isolate the term with the smallest coefficient in absolute value, namely $-4y$. Solving for $4y$, we obtain

$$4y = 6x + 5z - 2. \tag{2.12}$$

To reduce the coefficients, we add and subtract suitable multiples of 4. Specifically, consider

$$4u = 6x + 5z - 2 + (-4x - 4z) = 2x + z - 2,$$

which gives

$$z = 4u - 2x + 2.$$

We now express all three variables in terms of the free parameters x and u:

$$4y = 6x + 5(4u - 2x + 2) - 2 = -4x + 20u + 8,$$

hence

$$y = -x + 5u + 2.$$

Thus, all integer solutions of the equation (2.11) are

$$x = x,$$

$$y = -x + 5u + 2,$$

$$z = -2x + 4u + 2,$$

where $u, x \in Z$ are arbitrary integers. Hence, these are the general integer solutions to equation (2.11).

In some cases, additional conditions may be imposed on the solutions of equation (2.1). For example, suppose we require solutions to equation (2.11) such that the variable y is divisible by 7. Using the previous expression

$$y = -x + 5u + 2,$$

we impose the congruence condition

$$y \equiv 0 \ (mod\ 7),$$

i.e., there exists $v \in Z$ such that

$$7v = -x + 5u + 2.$$

Solving for x, we obtain

$$x = 5u - 7v + 2$$

Substituting this into the expression for y and z we get

$$y = 7v, z = -6u + 14v - 2.$$

Thus, all integer solutions satisfying the additional condition $7|y$ are given by:

For arbitrary u,$v \in Z$.

Another type of constraint imposed on equation (2.1) may take the form of an additional linear equation involving the same unknowns x, y, z. In such a case, we obtain a Diophantine system consisting of two linear equations with three unknowns. Since the number of equations is smaller than the number of variables, such a system, if consistent, admits infinitely many integer solutions.

Consider the general form of such a system:

$$ax + by + cz = d, \quad a_1x + b_1y + c_1z = d_1$$

where $a, b, c, d, a_1, b_1, c_1, d_1 \in Z$.

The conditions for the solvability of such a system are significantly more complex than in the case of a single linear Diophantine equation. Due to time and space constraints, we will not derive the general criterion here. Instead, we will demonstrate the solution method using a concrete numerical example.

Example 2. Find all integer solutions of the system of Diophantine equations:

$$x + 2y - 3z = 3, 3x + y + 2z = 1 \tag{2.13}$$

Solution. We begin by choosing a variable that appears in both equations. Here, we select the variable y. (Note: If no variable occurred in both equations, the procedure would be different.)

From the first equation we obtain:

$$2y = -x + 3z + 3,$$

From the second equation:

$$y = -3x - 2y + 1 \Rightarrow 2y = -6x - 4z + 2.$$

Now equating the two expressions for $2y$:

$$-x + 3z + 3 = -x - 4z + 2$$

Rewriting:

$$5x + 7z = -1 \tag{2.14}$$

Thus, the system is reduced to a single linear Diophantine equation in two variables.

Note: If the left-hand side of one equation is not a scalar multiple of the left-hand side of the other, not all variables are eliminated, and the system reduces to a single linear Diophantine equation with two unknowns, which can be solved using standard techniques.

Since $d(5,7) = 1$, equation (2.14) has integer solutions.

We solve for x:

$$5x = -1 - 7y \rightarrow \mathrm{x} = \frac{-1-7y}{5}.$$

To find integer values, we impose that the numerator must be divisible by 5. Let us express this condition in terms of an integer parameter. Let $u \in Z$ be such that:

$$5u = -1 - 7z + (5 + 10z) = 4 + 3z \rightarrow 3z = 5u - 4$$

Thus, $3|(5u - 4)$ which implies that there exists $v \in Z$ such that:

$$3v = 5u - 4 + (-6u + 3) = -u - 1 \rightarrow u = -1 - 3v$$
$$\Rightarrow 3z = 5(-1 - 3v) - 4 = -9 - 3.5v$$
$$\Rightarrow z = -3 - 5v \rightarrow 5x = -1 - 7(-3 - 5v) = 20 + 5.7v$$
$$\Rightarrow x = 4 + 7v$$
$$\Rightarrow y = -3(4 + 7v) - 2(-3 - 5v) + 1 = -5 - 11v.$$

Therefore, the general solution of the system (2.13) is:

$$x = 4 + 7v, \;\; y = -5 - 11v, \;\; z = -3 - 5v, v \in Z.$$

Verification:
Substitute the expressions into the original system:
1st equation:

$$x + 2y - 3y = 4 + 7v + 2(-5 - 11v| - 3(-3 - 5v) = 3$$
$$3x + y + 2z = 3(4 + 7v) - 5 - 11v + 2(-3 - 5v) = 1.$$

Notes:

The method for solving a linear Diophantine equation with three unknowns can be naturally extended to equations involving an arbitrary number of unknowns. The condition for the existence of integer solutions remains analogous: it depends on whether the greatest common divisor (GCD) of all coefficients divides the absolute term. If it does, integer solutions exist.

Similarly, systems of linear Diophantine equations with more unknowns can be solved in a manner analogous to the method used for a system of two equations in three variables. By successively eliminating unknowns, the system is gradually reduced to one or more equations with fewer variables, which can then be solved using known procedures.

In cases where we are required to find integer solutions of an equation (or a system of equations) involving rational coefficients, we first transform the equation(s) into an equivalent form with integer coefficients. This is done by multiplying both sides of each equation by the least common denominator (LCD) of all fractional coefficients. After this transformation, the resulting equations can be handled using the established methods for linear Diophantine equations.

Exercises:

(a) The difference between two two-digit numbers, each formed by reversing the digits of the other, is divisible by seven. Determine all such numbers.

(b) The ratio of two distinct three-digit numbers, each obtained by reversing the digits of the other, is divisible by seven. Identify these numbers.

Find an integer solution to the linear equation

$$5x + 6y - 8z = 92,$$

in which the variables x, y, z are numerically as close to one another as possible.

In the plane defined by the equation

$$3x + 4y - 7z = 1,$$

find the lattice point (i.e., a point with integer coordinates) whose coordinates are as close as possible to one another.

There are three types of containers available, each capable of storing 60, 80, or 150 boxes. How many containers of each type are needed to store exactly 10,000 boxes, assuming that the number of medium-sized containers (with a capacity of 80 boxes) must be minimized?

Find all integer solutions to the following system of linear equations:

$$2x - 3y + 5z = 4,$$

$$6x + 5y - 7z = 1.$$

A group of trainees — fewer than 100 in total — began by stepping in groups of eight, with three left over. When stepping in groups of five, two were left over, and in groups of six, one was left over. How many trainees were there?

Find all integer solutions to the system of equations:

$$x - \frac{2}{3}y + \frac{1}{2}z - 2u = 1,$$
$$3x + 2y - \frac{3}{4}z + \frac{1}{3}u = \frac{5}{2}.$$

Is it possible to fill 15 containers — each with a volume of 5, 6, or 8 liters — using exactly 100 liters (i.e., one hectoliter) of oil?

The total number of points earned in all 12 matches of a betting week was 21. How many outcomes of type "one" (worth 1 point), "two" (2 points), and "cross" (3 points) were there, given that each match results in exactly one of these outcomes?

Solutions:

(a) Let the original two-digit number be of the form $10x + y$, where $x, y \in N, 1 \leq x \leq 9, 1 \leq y \leq 9$ and $, x \geq y$ (Note: x and y must be non-zero, as they represent the tens and units digits of a valid two-digit number.) The reversed number is$10x + y$. According to the problem condition, the difference between these two numbers must be divisible by 7:

$$10x + y - (10y + x) = 7u, \text{ where } u \in Z \text{ or } 9(x - y) = 7u.$$

Thus, we require that $x - y$ satisfies:

$-8 \leq x - y \leq 8$, with $x > y \Rightarrow 0 < x - y \leq 8$

Additionally, since

$9(x - y) = 7u$, and $7u \leq 72$, it follows that $u - 9$.

This gives $x - y = 7$.
Possible pairs satisfying this condition are:

$x = 9, \ \ y = 2$ and $x = 8, \ \ y = 1$.

Thus, the searched numbers are either $92, 29$ or $81, 18$.

(b) Let the original three-digit number be of the form $100x + 10y + z$, where

$x, y, z \in N, 1 \leq x \leq 9, 0 \leq y \leq 9, 1 \leq z \leq 9$, and $x > z$.

The reversed number is $100z + 10y + x$. The condition implies that:

$100x + 10y + z \ - (100z + 10y + x) = 7u, u \in Z.$

So

$99(x - z) = 7u$

Therefore:

$$x - z = \frac{7u}{99}$$

We want $x - z$ to be a positive integer between 1 and 8. For $x - z$ to be an integer, $\frac{7u}{99}$ must also be an integer, i.e., $99|7u \Rightarrow 99|$u.
Let $u = 99$, then:

$$x - z = \frac{7.99}{99} = 7$$

Possible digit pairs satisfying $x - z = 7$ are:

$x = 9, z = 2,$
$x = 8, z = 1.$

Since the middle digit y can be any digit from 0 to 9, we obtain 10 values of y for each pair of (x, z) giving a total of:

2x10=20 valid three-digit number pairs.

Consider the Diophantine equation:

$5x + 6y - 8z = 92..$

Since the greatest common divisor of the coefficients satisfies $gcd\ (5,6,-8) = 1$, and $1 \mid 92$, the equation admits integer solutions.
Rewriting the equation in terms of x:

$$5x = 92 - 6y + 8z. \tag{2.15}$$

We aim to express the right-hand side in terms of parameters. Let us introduce an integer parameter $u \in Z$ and define:

$$5u = 92 - 6y + 8z - (90 - 5y + 5z) = 2 - y + 3z,$$

which leads to:

$y = 2 + 3z - 5u.$

Substituting this expression for y back into Equation (2.15), we obtain:

$5x = 92 - 6(2 + 3z - 5u) + 8z = 80 - 10z + 30u,$

which yields:

$x = 16 - 2z + 6u.$

Hence, the general solution to the given equation is:

$x = 16 - 2z + 6u, \ \ y = 2 + 3z - 5u, \ \ z = z,$

For all $z, u \in Z$.

Verification: $5(16 - 2z + 6u) + 6(2 + 3z - 5u) - 8z = 92$, confirms the correctness of the parametrization.

To determine the integer solution (x, y, z) in which the variables are numerically closest to one another—i.e., where the values $| x - y |, | x - z |$ and $| y - z |$ are minimized—we analyze the general parametric solution:

$x = 16 - 2z + 6u, \ \ y = 2 + 3z - 5u, \ \ z = z$, for $u, z \in Z$.

Let us denote the absolute differences:

$| x - y | = | 14 - 5z + 11u |, | x - z | = | 16 - 3z + 6u |, | y - z | = | 2 + 2z - 5u |$.

This formulation, however, is not optimal for minimizing all three differences simultaneously. Instead, we consider a reformulation of the original equation:

$5x + 6y - 8z = 92.$

Subtracting 5z+6z=11z from both sides gives:

$5(x - z) + 6(y - z) + 3z = 92.$

Let $a = x - z$ and $b = y - z$. Then this equation becomes:

$5a + 6b + 3z = 92.$

We now seek integer triples (a, b, z) for which the differences a, b, and 0 (i.e., $x - z, y - z$, and $z - z$) are small in absolute value.

Note that if $a = 0$ and $b = 0$, then:

$3z = 92 \Rightarrow z = \frac{92}{3},$

which is not an integer. Thus, a completely equal solution $x = y = z$ is impossible.

We now explore near-equal values:

Case 1:

Let $a = 1, b = 0 \Rightarrow x = z + 1, y = z - 1$.

Then:

$5(1) + 6(0) + 3z = 92 \Rightarrow 5 + 3z$ =92⇒z=29.

Therefore:

$x = 30, y = 29, z = 29$.

Case 2:

Let $a = 1, b = 1 \Rightarrow x = z + 1, y = z + 1$

Then:

$5(1) + 6(1) + 3z = 92 \Rightarrow 11 + 3z = 92 \Rightarrow z = 27$.

Thus:

$x = 28, y = 28, z = 27$.

Case 3:

Let $a = 1, b = -1 \Rightarrow x = z + 1, y = z - 1$.

Then:

$5(1) + 6(-1) + 3z = 92 \Rightarrow -1 + 3z = 92 \Rightarrow z = 31$.

Therefore:

$x = 32, y = 30, z = 31$.

Comparing the three cases, the second one minimizes all differences:

$$| x - y |= 0, | x - z |= 1, | y - z |= 1, |$$

which yields the smallest maximum absolute difference among the variables.

Conclusion:

The integer solution $(x, y, z) = (28,28,27)$ is optimal under the given condition that the values of the variables should be as close as possible.

$3x + 4y - 7z = 1$. Since $gcd\ (3,4,-7) = 11$, the given equation is solvable over the integers. We rewrite the equation as:

$$3x = 1 - 4y + 7z,$$

from which it follows that there exists $u \in Z$ such that:

$$3u = 1 - 4y + 7z + (3y - 6z) = 1 - y + z.$$

Hence,

$$y = 1 + z - 3u,$$

and subsequently:

$$3x = 1 - 4(1 + z - 3u) + 7z = -3 + 6z + 3u,$$

which implies:

$$x = -1 + 2z + u.$$

Therefore, all integer solutions of the given equation can be expressed in the parametric form:

$$x = -1 + 2z + u, y = 1 + z - 3u, z = z,$$

where $u, z \in Z$ are arbitrary integers.

According to the problem statement, we are interested in a solution that minimizes the values of $| x - z |, | x - y |$, and $| y - z |$. We reconsider the original equation, transforming it as follows:

$$3(x - y) + 4(y - z) = 1.$$

It follows that the case $x - y = 0, y - z = 0$ cannot occur. A candidate integer solution $x, y, z \in Z$ that fits the equation may satisfy:

$$y - z = 1, x - z = -1,$$

which yields:

$$1 - 3u = 1, -1 + z + u = -1.$$

This system has an integer solution only when $u = 0, z = 0$, implying that the sought lattice point is:

$$(-1,1,0).$$

Let x, y, and z denote the number of containers of the first, second, and third types, respectively. Then the following equation must be satisfied:

$$60x + 80y + 150z = 10000, x \geq 0, y \geq 0, z \geq 0. \quad (2.16)$$

We observe that:

$$gcd\ (60,80,150) = 10 \text{ and } 10 \mid 10000.$$

This implies that Equation (2.16) has an integer solution. Dividing both sides by 10 yields:

$$6x + 8y + 15z = 1000,$$

thus

$$6x = 1000 - 8y - 15z. \quad (2.17)$$

It follows that there exists $u \in Z$ such that

$$6u = 1000 - 8y - 15z - (996 - 6y - 12z) = 4 - 2y - 3z,$$

and therefore:

$$2y = 4 - 3z - 6u. \tag{2.18}$$

This implies that there exists $v \in Z$ such that:

$$2v = 4 - 3z - 6u + (-4 + 4z + 6u) = z.$$

Hence,

$$z = 2v \tag{2.19}$$

Substituting Equation (2.19) into Equation (2.18), we obtain:

$$2y = 4 - 3.2v - 6u.$$

and thus:

$$y = 2 - 3v - 3u. \tag{2.20}$$

Substituting Equation (2.19) and (2.20) into Equation (2.17) we get

$$6x = 1000 - 8(2 - 3v - 3u) - 15.2v = 984 + 24u - 6v.$$

So that:

$$x = 164 + 4u - v. \tag{2.21}$$

All solutions of Equation (2.16) are

$$x = 164 + 4u - v$$

$$y = 2 - 3v - 3u$$

$$z = 2v,$$

where $u, v \in Z$.

Moreover, since $y = 2 - 3v - 3u$, it follows that y is minimized when $u + v$ is maximized under the constraint $u + v \leq \frac{2}{3}$. The minimum occurs if and only if:

$$u + v = 0 \text{ or } u = -v.$$

Substituting $u = -v$ into the parametric form, we obtain the particular solution:

$$x = 164 - 5v, y = 2, z = 2v,$$

where $v \in Z$ and

$$v \geq 0, 164 - 5.v \geq 0$$

The latest inequality implies:

$$v \leq \frac{164}{5} = 32\frac{4}{5},$$

So, the valid integer values of v are:

$$v \in \{0,1,2, \ldots ,32\}.$$

Conclusion:

The Diophantine analysis demonstrates that the packing problem admits 33 distinct integer solutions, with the number of configurations determined by the range of admissible values for the parameter v. In each feasible solution, the number of medium containers remains constant at two, whereas the counts of small and large containers vary accordingly. This provides a complete parametric description of all optimal packing combinations under the given constraints.

Consider the system:

$$2x - 3y + 5z = 4,$$

$$6x + 5y - 7z = 1.$$

From the first equation, we have:

$$2x = 4 + 3y - 5z$$
$$\Rightarrow 6x = 12 + 9y - 15z,$$

from the second equation:

$$6x = 1 - 5y + 7z.$$

Equating both expressions for $6x$, we obtain:

$$12 + 9y - 15z = 1 - 5y + 7z \Rightarrow -14y + 22z = 11.$$

Since $gcd(-14,22) = 2$ and $2 \nmid 11$ (i.e., 2 does not divide 11), this Diophantine equation has no integer solution.

Therefore, the system has no integer solution.

Let us now denote x as the number of complete 8-step repetitions, y as the number of complete 5-step repetitions, and z as the number of complete 6-step repetitions. Then for the number of participants c, we must have:

$c = 8x + 3, c = 5y + 2, c = 6z + 1$, while $0 < c < 100$ is an integer.

We equate the expressions:

$$8x + 3 = 5y + 2, 8x + 3 = 6z + 1, \Rightarrow 5y + 2 = 6z + 1.$$

Rewriting:

$$5y - 6z = -1.$$

Since $gcd(5, -6) = 1$ and $1|(-1)$, the solution is solvable. From:

$$5y = 6z - 1 \Rightarrow 5|(6z - 1),$$

there exists $m \in Z$ such that:

$5m = 6z - 1 - 5z = z - 1 \Rightarrow z = 5m + 1.$

Substituting into the equation $5y = 6z - 1$, we get:

$5y = 6(5m + 1) - 1 = 30m + 5 \Rightarrow y = 6m + 1.$

From $c = 8x + 3 = 5y + 2 = 6z + 1$, we substitute $y = 6m + 1$ and $z = 5m + 1$ into:

$8x + 3 = 5(6m + 1) + 2 = 30m + 5 + 2 = 30m + 7 \Rightarrow 8x = 30m + 4 \Rightarrow 4x = 15m + 2.$

Now, there exists $p \in Z$ such that:

$4p = 15m + 2 \Rightarrow m = \frac{4p-2}{15}.$

Since we need integer solutions, we test values that lead to integer x, y, z, and compute:

$x = 8 - 15p, y = 13 - 24p, z = 11 - 20p$

Subject to the constraints $x > 0, y > 0, z > 0$, and:

$c = 8x + 3 = 67 - 120p, 0 < c < 100 \Rightarrow -\frac{33}{120} < p < \frac{67}{120}.$.

This holds only for $p = 0$, which satisfies all inequalities. Hence, the number of participants is:

$c = 67.$

Now consider the system:

$$x - \frac{2}{3}y + \frac{1}{2}z - 2u = 1,$$
$$3x + 2y - \frac{3}{4}z + \frac{1}{3}u = \frac{5}{2}.$$

Multiply the first equation by 6 and the second equation by 12:

$$6x - 4y + 3z - 12u = 6,$$
$$36x + 24y - 9z + 4u = 30.$$

From the first equation:

$$12u = 6x - 4y + 3z - 6$$

and from the second:

$$4u = 30 - 36x - 24y + 9z \Rightarrow \; 12u = 90 - 108x - 72y + 27z.$$

Therefore, it must be: $6x - 4y + 3z - 6 = \; 90 - 108x - 72y + 27z,$

Equating both expressions for $12u$:

$$6x - 4y + 3z - 6 = 90 - 108x - 72y + 27z \Rightarrow 114x + 68y - 24z = 96.$$

Simplifying:

$$57x + 34y - 12z = 48 \Rightarrow 12z = 57x + 34y - 48.$$

Then:

$$12 \mid (57x + 34y - 48).$$

Let $v \in \; Z$ such that:

$$12v = 57x + 34y - 48 + (48 - 48x - 36y) = 9x - 2y.$$

Then:

$$9x = 2y + 12v \Rightarrow x = 2t, y = 9t - 6v.$$

Substitute into:

$$12z = 57x + 34y - 48 = 420t - 204v - 48 \Rightarrow$$
$$z = 35t - 17v - 4.$$

From earlier, using the expression for $12u$:

$$12u = 81t - 27v - 18 \Rightarrow u = \frac{81t-27v-18}{12}.$$

To ensure $u \in Z$, we enforce:

$$4u = 27t - 9v - 6 \Rightarrow \exists m \in Z: 4m = -t - v - 2 \Rightarrow$$
$$v = -2 - t - 4m.$$

Then:

$$x = 2t, y = 12 + 15t + 24m, z = 30 + 52t + 68m,$$
$$u = 3 + 9t + 9m.$$

Hence, the full solution set is:

$$x = 2t,$$
$$y = 12 + 15t + 24m,$$
$$z = 30 + 52t + 68m,$$
$$u = 3 + 9t + 9m,$$
$$t, m \in Z$$

Let's denote the number of containers with 5, 6, and 8 liters by x, y and z, respectively. Given:

$$x + y + z = 15,$$
$$5x + 6y + 8z = 100.$$

From the first equation, after multiplying by the number 5, we get:

$$5x = 75 - 5y - 5z,$$

and from the second:

$$5x = 100 - 6y - 8z.$$

Equating:

$75 - 5y - 5z = 100 - 6y - 8z$, so $y + 3z = 25$.

As $gcd(1,3) = 1\ 1|25$, an integer solution exists. One such solution is $y_0 = 1, z_0 = 8$.

Letting $y - 1 + 3(z - 8) = 0$, and substituting into $x = 15 - y - z$

$x = 6 - 2m$.

Feasibility conditions:

$$x \geq 0 \Rightarrow m \leq 3, y \geq 0 \Rightarrow m \geq 0, z \geq 0 \Rightarrow m \leq 8.$$

m	0	1	2	3
x	6	4	2	0
y	1	4	7	10
z	8	7	6	5

Thus, $m \in \{0,1,2,3\}$ yielding four feasible solutions. If all three types of containers must be used, only the first three values of m qualify (i.e., $x, y, z > 0$).

9. Let x, y, z denote the number of ones, twos, and crosses, respectively. We are given the system:

$x + y + z = 12, x + 2y + 3z = 21$.

Subtracting the first equation from the second, we get:

$(x + 2y + 3z) - (x + y + z) = 21 - 12 \Rightarrow y + 2z = 9$.

One solution is $y_0 = 1, z_0 = 4$. Therefore, the general solution can be parametrized using an integer parameter $m \in Z$ as follows:

$y - 1 + 2(z - 4) = 0 \Rightarrow y - 1 = -2(z - 4) \Rightarrow y = 1 + 2m$,
$z = 4 - m$.

Substituting into the first equation:

$x = 12 - y - z = 12 - (1 + 2m) - (4 - m) = 7 - m.$

Now, to ensure all variables are non-negative integers:

- $x \geq 0 \Rightarrow 7 - m \geq 0 \Rightarrow m \leq 7$
- $y \geq 0 \Rightarrow 1 + 2m \geq 0 \Rightarrow m \geq -\frac{1}{2} \Rightarrow m \geq 0$ (since $m \in Z$)
- $z \geq 0 \Rightarrow 4 - m \geq 0 \Rightarrow m \leq 4$

Hence, $m \in \{0,1,2,3,4\}$, giving five integer solutions:

m	x	y	z
0	7	1	4
1	6	3	3
2	5	5	2
3	4	7	1
4	3	9	0

Remark:

The method of solving linear Diophantine equations with three variables can be readily generalized to equations with an arbitrary number of variables. The condition for solvability remains the same, namely, whether the greatest common divisor of the coefficients divides the constant term.

Let x, y, z denote the number of ones, twos, and crosses, respectively.

We are given the system:

$x + y + z = 12, x + 2y + 3z = 21.$

Subtracting the first equation from the second:

$(x + 2y + 3z) - (x + y + z) = 21 - 12 \Rightarrow y + 2z = 9$

This is a linear Diophantine equation in two variables. A particular solution is $y_0 = 1, z_0 = 4$, Therefore, the general solution can be written as:

$y = 1 + 2m, z = 4 - m$, for $m \in Z$.

Substituting into the first equation to find xxx:

$$x = 12 - y - z = 12 - (1 + 2m) - (4 - m) = 7 - m.$$

Now we determine the range of integer values for mmm such that $x, y, z \geq 0$:

- $x \geq 0 \Rightarrow 7 - m \geq 0 \Rightarrow m \leq 7,$
- $y \geq 0 \Rightarrow 1 + 2m \geq 0 \Rightarrow m \geq -\frac{1}{2},$
- $z \geq 0 \Rightarrow 4 - m \geq 0 \Rightarrow m \leq 4.$

Since mmm must be an integer, the valid values are $m = 0,1,2,3,4$
Thus, there are five valid solutions:

m	0	1	2	3	4
x	7	6	5	4	3
y	1	3	5	7	9
z	4	3	2	1	0

Remark:

The method used to solve linear Diophantine equations with three variables can be readily extended to equations with any number of variables. The condition for solvability is analogous: a solution exists if and only if the greatest common divisor (GCD) of the coefficients of the unknowns divides the constant term.

References

[1] Medvid', V. Algorithms of solving the Diophantine equation. In *ICERI 2023: 16th International Conference of Education, Research and Innovation, 13–15 November 2023, Seville, Spain* (conference proceedings). (2023).

[2] Rania, B. A. The mathematical analysis of linear Diophantine equations with two and three variables and its applications. *Egyptian International Journal of Engineering Sciences and Technology, November 2022*. (2022).

Chapter 3

Methods of Solving Some Diophantine Equations of the Second Degree

Abstract

In this chapter, we present methods for solving second-degree Diophantine equations. For each method, we identify the conditions for solvability and provide general algorithms. These algorithms are illustrated through specific examples to aid understanding.
This article was designed with high school students in mind, particularly those engaged in deeper and more intensive studies of mathematics and computer science. After mastering the methodology presented here, students demonstrated significantly improved abilities to solve these and related problems.
The author also emphasizes key educational goals intended for educators who wish to explore this subject further.
Educational Objectives:
This contribution serves as a resource to help teachers navigate the process of solving Diophantine equations. It aims to deepen and broaden their mathematical knowledge while enhancing critical thinking in this domain. Specifically, it enables teachers to:

- Conduct a thorough and accurate analysis of problems;
- Identify and implement effective problem-solving strategies;
- Synthesize various approaches to derive complete solutions.

When multiple solutions are possible, the article encourages the comparison of solution quality, complexity, and efficiency. This analytical approach helps teachers make informed decisions about the most appropriate and effective methods for particular types of problems. Ultimately, this process supports educators in drawing meaningful conclusions, comparing and classifying solution strategies, and choosing optimal approaches tailored to specific mathematical challenges.
All these educational aspects contribute to the development of teacher creativity, both in their teaching practice and in cultivating diverse problem-solving perspectives among students. The article also fosters flexibility in student thinking by encouraging a variety of solution strategies, which are beneficial for further study.

This material is recommended as a supplementary resource for both teachers and secondary school students. The more advanced problems included are especially suitable for students in specialized mathematics classes.

Keywords: Diophantine equation of the second degree, the Pythagorean equation, algorithm, condition, unknown, coefficient, the greatest common divisor

3.1. Introduction

Quadratic Diophantine equation of the second degree with two unknowns.

A quadratic Diophantine equation in two variables can take the form:

$$x^2 - y^2 = c, \tag{3.1}$$

where $c \in Z$ is a given integer. Our first goal is to determine the conditions under which this equation has integer solutions. See e.g., Dorin and Titu [1], Choudhry [2] or Medviď [3].

We begin by factoring the left-hand side of equation (3.1):

$$(x + y).(x - y) = c. \tag{3.2}$$

We now consider the parity (evenness or oddness) of x and y:

- If x and y are both even or both odd (i.e., have the same parity), then both $x + y$ and $x - y$ are even. Hence, c, being the product of two even numbers, must be divisible by 4.
- If x and y have different parity, then both $x + y$ and $y - x$ are odd, and therefore their product c must be odd.

Thus, for equation (3.2) to have integer solutions, the number c must either be divisible by 4 or be odd.

Equation (3.2) also provides a constructive method for solving equation (3.1). Given such a c, one must find a factorization of c into two integers of the same parity. Assign the larger factor to $x + y$ and the smaller to $x - y$, then solve for x and y:

$x = \frac{(x+y)+(x-y)}{2}$, $y = \frac{(x+y)-(x-y)}{2}$

A special case arises when both factors are equal, i.e., $x + y = x - y$, which implies $y = 0$ and $x^2 = c$. In this case, equation (3.1) has a solution if and only if c is a perfect square.

Therefore, if c is an odd number greater than 1 or a multiple of 4 greater than 4, then at least one such factorization exists (e.g., $c = 2n + 1 = 1 \cdot (2n + 1)$, or $c = 4n = 2 \cdot (2n)$) and the equation has at least one solution.

Conclusion:

We conclude that equation (3.1) has an integer solution if and only if c is either an odd integer or divisible by 4.

Example 1. Find an integer solutions to the following equations

a) $x^2 - y^2 = 21$

b) $x^2 - y^2 = 8$

c) $x^2 - y^2 = 18$

Solution.

a) We factor the right-hand side:

21 = 1.21 = 3.7.

Both pairs consist of integers of the same parity.

For $x + y = 21, x - y = 1$,.

$x = \frac{21+1}{2} = 11$, $y = \frac{21-1}{2} = 10$.

For $x + y = 7, x - y = 3$,

$x = \frac{7+3}{2} = 5$, $y = \frac{7-3}{2} = 2$.

Thus, solutions are $(x, y) = (11,10)$ and $(5,2)$.

b) Factorization:

$8 = 4.2 = 8.1$

Only one pair has the same parity.

For $x + y = 4, x - y = 2,$

Solving:

$x = \frac{4+2}{2} = 3, y = \frac{4-2}{2} = 1.$

Thus, one solution is $x_1 = 3, y_1 = 1.$

Second pair: $x + y = 8, x - y = 1,$

This is invalid, as 8 and 1 have different parity, so no integer solution arises from this decomposition.

Conclusion: Equation (b) has integer solutions, e.g., $(x, y) = (3,1)$

c) Sincee 18≡ 2 mod (4),

but $x^2 - y^2 = 0 \; or \; 1 \bmod(4),$

equation (c) has no integer solutions.

Comment. We presented only positive solutions. However, integer solutions with negative values of x and/or y are also possible.

3.2. Diophantine Bilinear Equation

The Diophantine bilinear equation is an equation of the form

$$a.x.y + b.x + c.y + d = 0, \tag{3.3}$$

where $a, b, c, d \in Z$, assuming that a, c are coprime integers and $a, b, c \neq 0$.

From (3.3) we can express

$$y = \frac{-bx-d}{ax+c}. \tag{3.4}$$

We seek integer values of x such that the right-hand side of (3.4) is an integer. Let us set

$$bx + d = u, ax + c = v. \tag{3.5}$$

Eliminating x from equations (3.5):

$$abx + ad = au,$$
$$abx + bc = bv.$$

Subtracting gives

$$au - bv = ad - bc. \tag{3.6}$$

If v is a divisor of u, then according to (3.6) v must also divide $a.d - b.c$.

From (3.6) it follows that:

$$au = (ad - bc) + bv,$$

If $v \mid u$ then from (3.6) it follows that $v \mid (ad - bc)$.

Since a and c are coprime by assumption, the second equation in (3.5) implies that a and v are also coprime. Therefore, from $v \mid au$ we conclude $v \mid u$.

Hence, any integer x satisfying $ax + c = v$ yields an integer solution of (3.4).

The numbers a, c are coprime numbers according to the assumption. The second equation of (3.5) shows that the numbers a, v are coprime numbers, because each common divisor of a, v is also the divisor of $a, c = v - a.x$. Since v is the divisor of $a.u$ and a, v are coprime numbers, thus $v \mid u$. The integer number x, which satisfies the equation $a.x + c = v$, thus leads to the solution of equation (3.4).

Comment. So far, we have assumed that a, c are coprime.

What happens if a, c are not coprime?

Suppose that $\delta = d(a,c) \neq 1$ be the greatest common divisor of a, c. Then we can rewrite (3.4) in the following form

$$y = \frac{-1}{\delta}\frac{bx+d}{\frac{a}{\delta}x+\frac{c}{\delta}}. \tag{3.7}$$

Let us define $a' = \frac{a}{\delta}$, $c' = \frac{c}{\delta}$, and consider the equation

$$y' = \frac{-(bx+d)}{a'x+c'}. \tag{3.8}$$

From all the values of y'that satisfy equation (3.8), we consider only those which are divisible by δ. For such values, we define $y = \frac{y'}{\delta}$.

Example 2. *Find all integer solutions of the following equations:*

a) $4.x.y - 3.x + 5.y + 19 = 0$
b) $16.x.y - 3.x + 20.y + 19 = 0$

Solution.

We rewrite the given equation in the form $y = \frac{3x-19}{4x+5}$.

In this case, we compute

$a.d - b.c = 4.19 - (-3).5 = 76 + 15 = 91.$

We find all divisors v of the number 91. Thus:

$v \in \{-1, +1, -7, +7, -13, +13, -91, +91\}.$

For $v = 1, -7, 13, -91$, we obtain $x = -1, -3, 2, -24$, leading to $y = -22, 4, -1, 1$.

Thus, solutions are $(-1,-22), (-3,4), (2,-1), (-24,1)$.

For the remaining values of v, no integer solution for x arises.

Here

$\delta = gcd(a,c) = gcd(16{,}20) = 4.$

Therefore, we rewrite the equation in the form:

$y = \frac{1}{4} \cdot \frac{3x-19}{4x+5},$

Auxiliary equation:

$y' = \frac{3x-19}{4x+5}.$

Using the results from part (a), we find that the given equation has only one solution satisfying the condition $y' \in Z$, namely:

$x = -3$ and $y = 1$.

Comment. In previous considerations we have silently assumed that $a.d - b.c \neq 0$.

What happens if $ad - bc = 0$?

If $ad - bc = 0$ then:

$d = \frac{bc}{a}.$

Substituting into equation (3.4) we obtain:

$y = \frac{-b.x+\frac{b.c}{a}}{a.x+c} = -\frac{b}{a}.$

Substituting $y = \frac{-b}{a}$ and $d = \frac{b.c}{a}$ back into (3.3) we get:

$ax.\left(\frac{-b}{a}\right) + bx + c.\left(\frac{-b}{a}\right) + \frac{bc}{a} = -bx + bx - \frac{bc}{a} + \frac{bc}{a} = 0.$

Thus, the equation is identically satisfied. However, we require integer solutions. Therefore, the condition:

$$a \mid b.$$

must hold. Thet is, there exists $k \in Z$such that $b = k.a$. Consequently, we obtain the general solution:

$$x \in Z, y = -k.$$

3.3. Pythagorean Equation

One of the most important simple Diophantine equations of the second degree is the equation

$$x^2 + y^2 = z^2. \tag{3.9}$$

The solutions of (3.9) in the domain of integers are called Pythagorean triples. Geometrically, they correspond to right-angled triangles with integer side lengths.

3.3.1. Primitive Solutions

We first determine only the primitive solutions of equation (3.9), i.e., those for which $gcd(x, y, z) = 1$. All other solutions are obtained by scaling.

If $\delta = gcd(x, y, z) > 1$, we can write

$$x = \delta x', y = \delta . y', z = \delta z',$$

where $x', y', z'$are coprime integers. Substituting into equation (3.9) and dividing by δ^2, we obtain

$$(x')^2 + (y')^2 = (z')^2,$$

showing that x', y', z' is also a solution of (3.9). with the required coprimality.

3.3.2. Parity Considerations

From this is follows that x and y cannot both be even. Otherwise, z^2 would also be even, and hence $gcd(x, y, z) \geq 2$, contradicting primitivity.

Similarly, x and y cannot both be odd. Indeed, if

$$x = 2\zeta + 1, y = 2\eta + 1$$

then

$$z^2 = x^2 + y^2 = 4(\zeta^2 + \eta^2 + \zeta + \eta) + 2,$$

which is even but not divisible by 4. This is impossible, since the square of an even integer is always divisible by 4.

Thus, z^2 would be even but not divisible by 4, which is impossible since squares of even numbers are divisible by 4.

Therefore, x and y must have different parity, and hence z is odd. Without loss of generality, we assume x is even and y is odd.

3.3.3. Factorization Method

Equation (3.9) can be rewritten as

$$(z + y)(z - y) = x^2 \tag{3.10}$$

Multiplying both sides by $\frac{1}{4}$, we obtain:

$$\frac{1}{2}(z + y).\frac{1}{2}(z - y) = \frac{1}{4}x^2. \tag{3.11}$$

Since z and y are odd, both $\frac{1}{2}(z + y)$ and $\frac{1}{2}(z - y)$ are integers. Moreover, they are coprime: if they had a common divisor $p > 1$, then then p would also divide both z and y, contradictoring $\gcd(x, y, z) = 1$.

Thus, $\frac{1}{2}.(z + y)$ and $\frac{1}{2}(z - y)$ are coprime integers whose product is a square. Therefore, each factor must itself be a perfect square. Let

$$\frac{1}{2}(z+y)=u^2, \frac{1}{2}(z-y)=v^2,$$

For some integers u, v. Then

$$\frac{1}{4}x^2=u^2v^2 \Rightarrow x=2uv.$$

From these relationships we derive the general parametric form of all primitive Pythagorean triples:

$$x=2uv, y=u^2-v^2, z=u^2+v^2,$$

where $u>v, \gcd(u,v)=1$ and u, v have opposite parity.

All solutions of equation (3.9) can are then given by

$$x=2kuv, y=k.(u^2-v^2), z=k.(u^2+v^2), \quad (3.12)$$

where $k \in N, u>v, u, v, \gcd(u,v)=1$ and u, v have different parity.

Examples:

For $u=2, v=1, k=1$: $(x,y,z)=(4,3,5)$, the classical triple.
For $u=3, v=2, k=1$: $(x,y,z)=(12,5,13)$.
For $u=4, v=3, k=1$: $(x,y,z)=(24,7,25)$,

and so on.

Example 3. *Find all integer solutions of equation*

$$x^2+4y^2=(x+y+z)^2.$$

Solution. We present two different approaches using different substitutions; both reduce the problem to the Pythagorean equation.

First approach

Set $x=Y, 2y=X, x+y+z=Z$,

Then the given equation becomes

$$X^2 + Y^2 = Z^2 ,$$

the Pythagorean equation. Its integer solutions admit the standard parametrization

$$X = 2kuv, Y = k(u^2 - v^2), Z = k(u^2 + v^2), \tag{3.13}$$

with $k \in N, u > v, \gcd(u, v) = 1$, and u, v of opposite parity. Reverting to the original variables gives

$$x = Y = k(u^2 - v^2), y = \frac{1}{2}.X = \frac{1}{2}.2kuv = kuv.$$

Finally,

$$z = Z - x - y = k(u^2 + v^2) - k(u^2 - v^2) - kuv = 2kv^2 - kuv = kv(2v - u),$$

So, one family of integer solutions is

$$x = k(u^2 - v^2), y = kuv, z = kv(2v - u),$$

with $k \in N, u > v, gcd(u, v) = 1$ and u, v of opposite parity.

(Observe that z may be negative for some choices of u, v; if one requires nonnegative side lengths, require $2v > u$)

Second approach:

Multiply both sides of the original equation by $\frac{1}{4}$ and set

$$\frac{x}{2} = X, y = Y, \frac{x+y+z}{2} = Z.$$

Then again

$$X^2 + Y^2 = Z^2,$$

so, we may use the same parametrization

$$X = 2kuv, Y = k(u^2 - v^2), Z = k(u^2 + v^2),$$

with the same constraints $k \in N, u > v, gcd(u, v) = 1$ and u, v of opposite parity.

Returning to x, y, z yields

$$x = 2X = 4kuv, y = Y = k(u^2 - v^2).$$

and

$$z = 2Z - x - y = 2k(u^2 + v^2) - 4kuv - k(u^2 - v^2) = k(u^2 - 4uv + 3v^2).$$

Hence the second family of integer solutions is

$$x = 4kuv, y = k(u2 - v^2), z = k(u^2 - 4uv + 3v^2)$$

with the same restrictions on k, u, v.

Remarks:

- Both approaches are correct and produce integer solution families; they differ because of different linear substitutions and therefore parametrize different subsets of solutions (they are not the same parametrization).
- If nonnegative z (or geometric triangle side-lengths) are required, impose the additional inequality constraints (for instance, in the first family require $2v \geq u$; in the second family require $u^2 - 4uv + 3v^2 \geq 0$).
- You may also combine both parametrizations to describe all integer solutions (allowing all integer values of k, u, v under the usual coprimality and parity conditions).

Example 4. *Find right-angled triangles with integer sides, whose longer leg is the arithmetic mean of shorter leg and hypotenuse.*

Solution.

We seek integer solutions to equation (3.9), described by the parametric relations (3.12), under the condition that the longer leg is the arithmetic mean of the other two sides. That is, we require

$$x = \frac{y+z}{2},$$

where x, y, and z are the sides of a right-angled triangle, with x as the longer leg.

Using the standard parametrization of primitive Pythagorean triples, we have

$$x = 2kuv, y = k(u^2 - v^2), z = k(u^2 + v^2),$$

where $k \in u > v, gcd\ (u, v) = 1$and u, v have opposite parity.

Substituting into the condition $x = \frac{y+z}{2}$, we obtain

$$2kuv = \frac{k(u^2-v^2)+k(u^2+v^2)}{2} = \frac{2ku^2}{2} = ku^2.$$

Cancelling k gives

$$2uv = u^2 \Rightarrow u = 2v.$$

Substitution

Putting $u = 2v$ into the parameterization yields

$$x = 2k(2v)(v) = 4kv^2,$$
$$y = k((2v)^2 - v^2) = 3kv^2,$$
$$z = k((2v)^2 + v^2) = 5kv^2.$$

Thus all such right-angled triangles have side lengths

$$(x, y, z) = (4kv^2, 3kv^2, 5kv^2),\ k, v \in N.$$

where $k = v^2 \in N$. These are precisely all integer-scaled versions of the classic Pythagorean triple (3,4,5).

Conclusion:

Up to scaling, these are exactly the integer multiples of the classical Pythagorean triple (3,4,5).

Example 5. *Identify all right-angled triangles with integer sides that have a given perimeter.*

Solution.

The side lengths of a right-angled triangle with integer values can be expressed means of the Pythagorean triple parameterization:

$$x = 2kuv, y = k(u^2 - v^2), z = k(u^2 + v^2),$$

where $k \in N, u > v, \gcd(u, v) = 1$ and u, v have opposite parity.

Let the perimeter be $x + y + z = 2s$, so that s is the semi-perimeter. Substituting the above parameterization, we obtain

$$x + y + z = 2kuv + k(u^2 - v^2) + k(u^2 + v^2) = 2s.$$

Simplification yields

$$2kuv + ku^2 = 2s \Rightarrow ku(u + v) = s.$$

Thus, for a given semi-perimeter s, one must find integers k, u, v satisfying the conditions above such that

$ku(u + v) = s, u > v, \gcd(u, v) = 1$ and u, v have opposite parity.

Example: Let the parameter be $2s = 112$ so $s = 56$.

Sinve

$$56 = 2^3.7 = 2.4.(4 + 3),$$

we may take $k = 2, u = 4, v = 3$.

Substituting into the formulas for (x, y, z) we obtain

$$\mathrm{x} = 2kuv = 2.2.4.3 = 48,$$

$y = k(u^2 - v^2) = 2(16 - 9) = 14,$
$z = k(u^2 + v^2) = 2(16 + 9) = 50.$

Hence $(x, y, z) = (48, 14, 50)$ is a Pythagorean triple with perimeter 112, which satisfies the required conditions.

Conclusion

In this work, we examined elementary quadratic Diophantine equations of the second degree with respect to their solvability. We began with the study of a quadratic equation in two variables, whose solution turned out to be relatively straightforward. The second case addressed a bilinear Diophantine equation, which required a more elaborate analysis.

The third and most substantial case—the Pythagorean equation—may justly be regarded as the *queen* among quadratic Diophantine equations. This classical equation not only possesses a rich theoretical structure but also exhibits numerous applications. The analysis of particular instances provides insights of lasting value, contributing not only to algebra but also to geometry and its practical implementations.

For each type of equation considered, we derived general conditions of solvability and presented the corresponding solutions. To clarify the theoretical results, we supplemented the discussion with concrete examples accompanied by complete derivations.

3.4. Algorithms of Solving the Diophantine Equation

$x^2 + y^2 = \alpha z^2$

We consider the Diophantine equation

$$x^2 + y^2 = \alpha z^2, \alpha \in Z.$$

and investigate its solvability for various integer values of α. Our aim is to determine the conditions under which solutions exist, and to derive explicit solution formulas whenever possible.

Case $\alpha = 1$: The Pythagorean Equation
In this case, we obtain the classical Pythagorean equation

$$x^2 + y^2 = z^2. \tag{3.14}$$

Its integer solutions are called Pythagorean triples. Geometrically, they correspond to right-angled triangles with integer side lengths.

e first restrict our attention to primitive solutions, i.e., triples (x, y, z) with $gcd\ (x, y, z) = 1$. All non-primitive solutions can then be obtained by multiplying primitive triples by a common factor. Indeed, if $\delta = gcd\ (x, y, z) > 1$, we may write

$$x = \delta x', y = \delta y'\ z = \delta z',$$

And substituting into (3.14), followed by δ^2, yields

$$(x')^2 + (y')^2 = (z')^2,$$

where (x', y', z') is primitive triple.

3.4.1. Parity Restrictions

From the assumption of coprimality, it follows that x and y cannot both be even; otherwise, z would also be even, and $gcd\ (x, y, z) \geq 2$. Similarly, x and y cannot both be odd. Indeed, if

$$x = 2\zeta + 1, y = 2\eta + 1,$$

Then

$$z^2 = x^2 + y^2 = 4(\zeta^2 + \eta^2 + \zeta + \eta) + 2.$$

which is even but not divisible by 4 - a contradiction. Therefore, one of x, y must be even, the other odd, and consequently z is odd.

Without loss of generality, assume that x is even and y is odd. Then equation (3.14) can be rearranged as

$$(z+y).(z-y) = x^2. \quad (3.15)$$

Dividing by 4 gives

$$\frac{1}{2}(z+y).\frac{1}{2}(z-y) = \frac{1}{4}x^2. \quad (3.16)$$

Since both z and y are odd, the quantities $\frac{1}{2}(z+y), \frac{1}{2}(z-y)$ are integers. Moreover, they are coprime; otherwise, their greatest common divisor would divise x, y, z simultaneously, contradicting coprimality.

3.4.2. Parametrization

Equation (3.16) therefore represents a product of two coprime integers that is a perfect square. Hence each factor must itself be a perfect square. Let

$$\frac{1}{2}(z+y) = u^2, \frac{1}{2}(z-y) = v^2, u > v,$$

with $gcd\,(x, y, z) = 1$ and u, v of opposite parity. Then

$$\frac{1}{4}x^2 = u^2v^2 = 2uv.$$

$$x = 2uv,\ y = u^2 - v^2,\ z = u^2 + v^2,$$

This yields the complete parametrization of primitive solutions.
For general (non-primitive) solutions, we multiply by a positive integer k:

$$x = 2kuv,\ y = k(u^2 - v^2),\ z = k(u^2 + v^2), \quad (3.17)$$

where $k \in N, u > v, gcd(u, v) = 1$ and u, v have different parity.

Examples

$u = 2, v = 1, k = 1 \Rightarrow (x, y, z) = (4, 3, 5),$
$u = 3, v = 2, k = 1 \Rightarrow (x, y, z) = (12, 5, 13),$
$u = 4, v = 3, k = 1 \Rightarrow (x, y, z) = (24, 7, 25),$

and so on.

Thus, all integer solutions of the Pythagorean equation (3.14) are generated by formula (3.17).

Diophantine equation $x^2 + y^2 = \alpha z^2$,

Case $\alpha = 0$

This equation is

$$x^2 + y^2 = 0, \tag{3.18}$$

Holds if and only if $(x, y, z) = (0, 0, z), z \in Z$.

Case $\alpha < 0$

Let α be a negative integer. The equation

$$x^2 + y^2 = \alpha z^2,$$

has only the trivial solution $(x, y, z) = (0, 0,0)$.

Indeed, the left-hand side, $x^2 + y^2$ is non-negative for all integer values of x and y, while the right-hand side, αz^2 is strictly negative for any $z \neq 0$ because $\alpha < 0$. Hence z must be 0; substituting $z = 0$ gives $x^2 + y^2 = 0$, so $x = y = z = 0$.

Diophantine equation $x^2 + y^2 = \alpha z^2$,

Case $\alpha = 2$

We now examine the equation

$$x^2 + y^2 = 2z^2. \tag{3.19}$$

Trivial solution.

One obvious solution is the trivial solution:

$(x, y, z) = (\pm k, \pm k, \pm k), k \in Z.$

Non-trivial Solutions

We now seek other, non-trivial solutions in the form:

$x = k + l, y = k - n, z = k$, where $k, l, n \in Z.$

Substituting into (3.19), we get

$$(k + l)^2 + (k - n)^2 = 2k^2. \tag{3.20}$$

Let us expand the left-hand side:

$(k + l)^2 + (k - n)^2 = k^2 + 2.k.l + l^2 + k^2 - 2.k.n + n^2 = k^2 + k^2 + l^2 + n^2 + 2k(l - n) = 2.k^2.$

Thus, equation (3.20) becomes:

$$2k^2 + l^2 + n^2 + 2k(l - n) = 2.k^2.$$

Which simplifies to:

$$l^2 + n^2 + 2.k.l - 2.k.n = 0.$$

This is a quadratic equation in n, with integer parameters k and l. To solve it, we treat it as a quadratic in n and apply the quadratic formula. Rearranging:

$$n^2 - 2kn + (l^2 + 2kl) = 0,$$

and solving gives:

$$n_{1,2} = k \pm \sqrt{(k^2 - l^2 - 2.k.l)},$$

Let:

$$D = (k^2 - l^2 - 2.k.l).$$

For n to be an integer, the discriminant D must be a perfect square. So we require:

$$k^2 - l^2 - 2kl = a^2 \tag{3.21}$$

where $a \in Z$.

Equation (3.21) can be rewritten as:

$$l^2 + 2kl + a^2 - k^2 = 0, \tag{3.22}$$

which is a quadratic equation in l, now with parameters k and a. Solving for l:

$$l_{1,2} = -k \pm \sqrt{(2k^2 - a^2)}.$$

For this solution to yield integer values of l, we again require the discriminant to be a square. That is:

$2k^2 - a^2 = b^2$, where $b \in Z$.

3.4.3. Algorithmic Approach

The question now becomes: how can we choose integer parameters k and a so that $D = 2k^2 - a^2$ is a perfect square?

We propose the following algorithmic method:

1. Fix an integer value of k,
2. Iterate through integer values $a < k$,
3. For each a compute $D = 2k^2 - a^2$,
4. Check whether D is a perfect square.

This procedure is illustrated, which presents computed values for various choices of k and a, highlighting cases where the discriminant becomes a square, hence yielding valid integer solutions for l and n; see Table 1.

Table 1. Integer solutions of equation (3.22)

k	1	2	3	3	4	4	4	5	5	5	5
a	0	1	1	2	1	2	3	1	2	3	4
$2k^2 - a^2$	2	7	17	14	31	28	23	49	46	41	34

k	6	6	6	6	6	7	7	7	7	7	7
a	1	2	3	4	5	1	2	3	4	5	6
$2k^2 - a^2$	71	68	63	56	47	97	94	89	85	73	62

k	8	8	8	8	8	8	8
a	1	2	3	4	5	6	7
$2k^2 - a^2$	127	124	119	112	103	92	79

k	9	9	9	9	9	9	9	9
a	1	2	3	4	5	6	7	8
$2k^2 - a^2$	161	158	153	146	137	122	113	98

k	10	10	10	10	10	10	10	10	10
a	1	2	3	4	5	6	7	8	9
$2k^2 - a^2$	199	196	191	184	175	164	151	136	119

When is $D = 2k^2 - a^2$ a Perfect Square?

Clearly, if k is an integer multiple of a, i.e., there exists $\omega \in Z$ such that $k = \omega a$, then:

$$D = 2(\omega a)^2 - a^2 = a^2(2\omega^2 - 1),$$

and thus the expression $2\omega^2 - 1$ must be a perfect square.

Table 2 represents the evaluation of the expression $2\omega^2 - 1$.

Table 2. Integer solution of equation (3.22)

ω	1	2	3	4	5
$2\omega^2 - 1$	1	7	17	31	49

From Table 2 we observe that the expression $2\omega^2 - 1$ is a perfect square (highlighted in bold) for $\omega = 1$ and $\omega = 5$. Through gradual calculation, it becomes evident that for no other natural number ω is the expression $2\omega^2 - 1$ a square.

If $\omega = 1$, then $k = a$. Substituting into equation (3.22) , we obtain

$l_1 = 0$ and $l_2 = -2k$,

and consequently

$n_1 = 0$ and $n_2 = 2k$.

Substituting into equation (3.20) we get

$$(\pm k)^2 + (\pm k)^2 = 2.k^2, \tag{3.23}$$

where $x = \pm k, y = \pm k, z = \pm k, k \in Z$. This corresponds to the known trivial solution.

If $\omega = 5$ then $k = 5a$. Substituting into equation (3.22), we obtain:

$l_{1,2} = -k \pm \sqrt{\left(k^2 + k^2 - \frac{k^2}{25}\right)} = -k \pm \frac{7k}{5}$, i.e., $l_1 = \frac{2}{5}k$ and $l_2 = -\frac{12}{5}k$.

Consequently

$$n_{1,2} = k \pm \sqrt{\left(k^2 - \left(\frac{2}{5}k\right)^2 - 2k\left(-\frac{2}{5}k\right)\right)} =$$

$$k \pm \sqrt{\left(\frac{25k^2 - 144k^2 + 120k^2}{25}\right)} = k \pm \frac{k}{5}.$$

Hence:

$n_1 = \frac{6}{5}k$ and $n_2 = \frac{4}{5}k$.

Substituting l_1, n_1, l_2, n_2 into equation (3.20) and obtain

$$(\pm\tfrac{7}{5}k)^2 + (\pm\tfrac{1}{5}k)^2 = 2k^2. \tag{3.24}$$

Since $k = 5.a$ thus the number k gains the natural multiples of the number 5, i.e., $k \in \{5,10,15,20, \dots\}$.

Consequently, the triple $x = \pm\frac{7}{5}k, y = \pm\frac{1}{5}k, z = k$ is the another solution for our equation. Substituting $\frac{k}{5} = \varepsilon$ then we obtain $\varepsilon \in N$, where N is the set of natural numbers.

Thus our solution will be the triple $x = \pm 7\varepsilon, y = \pm\varepsilon, z = 5\varepsilon, \varepsilon \in N$.

Solution 2.

It is clear that if (x, y, z) is an integer solution of the given equation, then the quantity $x^2 + y^2$ must be divisible by 2. Since any integer $c \in Z$ can be expressed in exactly one of the forms $c = q$ or $c = 2q + 1$, where $q \in Z$, we find that:

$$c^2 = \begin{Bmatrix} 4q^2 \; if \; c = 2q \\ 4q^2 + 4q + 1 \; if \; c = 2q + 1 \end{Bmatrix}$$

This implies that the square of an integer modulo 2 yields a remainder of either 0 or 1. Consequently, the possible remainder classes for x^2, y^2, and $x^2 + y^2$ modulo 2 are summarized in the next table; see Table 3.

Table 3. Remainder Class Modulo 2

	Remainder class	Remainder class	Remainder class	Remainder class
x^2	0	0	1	1
y^2	0	1	0	1
$x^2 + y^2$	0	1	1	0

Given this analysis, we now examine the equation:

$$(2x_1 + 1)^2 + (2y_1 + 1)^2 = 2z^2, \tag{3.25}$$

where we define $2x_1 + 1 = x$ and $2y_1 + 1 = y$, ensuring that both x and y are odd integers.

The results of this computation are summarized in in the next table; see Table 4.

Table 4. Values for equation (3.25))

x_1	y_1	x	y	$x^2 + y^2 = 2.z^2$	z
0	0	1	1	1+1=2.1	1
1	0	3	1	9+1=2.5	
2	0	5	1	25+1=2.23	
2	1	5	3	25+9=2.17	
3	0	7	1	49+1=2.5^2	5
3	1	7	3	49+9=2.29	
3	2	7	5	49+25=2.37	
4	0	9	1	81+1=2.41	

Note: Rows in bold indicate valid integer solutions for the equation.

Thus, we find two classes of solutions:

1. The trivial solution: $(x, y, z) = (k, k, k)$, where $k \in Z$
2. The nontrivial solution: $(x, y, z) = (7k, k, 5k)$, where $k \in Z$.

These are precisely the same solutions obtained in Solution 1, derived here through a different modular and parity-based argument.

Remark: Only a portion of the table is presented here, as further values of $x_1$and y_1 do not yield additional integer solutions for x, y, z satisfying the equation.

Diophantine Equation $x^2 + y^2 = \alpha z^2$

Case $\alpha = 3$

Let us consider the Diophantine equation:

$$x^2 + y^2 = 3z^2, \tag{3.26*}$$

It is clear that if (x, y, z) is a solution of this equation, then the sum $x^2 + y^2$ must be divisible by 3.

Every integer $c \in Z$ can be written in exactly one of the forms

$$c = 3q, c = 3q + 1, c = 3q - 1,$$

where $q \in Z$. Squaring each form gives:

$$c^2 = 9q^2, c^2 = 9q^2 + 6q + 1 \text{ or } c^2 = 9q^2 - 6q + 1,$$

respectively.

Thus, the square of an integer modulo 3 always belongs to the residue class 0 or 1. Consequently, the possible residue classes of x^2, y^2, and $x^2 + y^2$ modulo 3 are summarized in in the next table; see Table 5.

Table 5. Remainder Classes Modulo 3

	Remainder class	Remainder class	Remainder class	Remainder class
x^2	0	0	1	1
y^2	0	1	0	1
$x^2 + y^2$	0	1	1	2

From the last row of Table 5, we observe that for $x^2 + y^2$ to be divisible by 3, both x and y must be divisible by 3. That is, we may write $x = 3x_1$ and $y = 3y_1$, where $x_1, y_1 \in Z$.

Substituting into the equation (3.26 $*$) yields

$$(3x_1)^2 + (3y_1)^2 = 3.z^2.$$

which simplifies to

$$3(x_1^2 + y_1^2) = z^2. \tag{3.26}$$

This shows that z^2 is divisible by 3, and hence zzz itself must also be divisible by 3. Writing $z = 3z_1$ with $z_1 \in Z$, equation (3.26) becomes

$$3(x_1^2 + y_1^2) = 9z_1^2 \Rightarrow x_1^2 + y_1^2 = 3z_1^2.$$

Thus, starting from a solution (x, y, z), we have obtain a smaller solution (x_1, y_1, z_1) of the same type. If $z > 0$ is minimal among all solutions, this process gives a contradiction, since it produces a strictly smaller positive integer.$z_1 < z$.

$$(x, y, z) = (0,0,0).$$

Diophantine Equation $x^2 + y^2 = \alpha z^2$

Case $\alpha = 4$

We now examine the Diophantine equation

$$x^2 + y^2 = 4z^2$$

This can be rewritten as

$$x^2 + y^2 = (2z)^2. \tag{3.27}$$

Letting $w = 2z$, we transform the equation into:

$$x^2 + y^2 = w^2, w \in Z^+$$

Equation (3.27) is a Pythagorean equation, already discussed in Section 1 (cf. Equation 3.14), with the additional condition that the hypotenuse w must be even. Consequently, both x and y must also be even, and w^2must be divisible by 4. This requirement is trivially satisfied, since $w = 2z$ implies

$$w^2 = 4z^2 \Rightarrow 4 \mid w^2.$$

From our earlier results on Pythagorean triples, all solutions to Equation (3.14) are given by

$$x = 2kuv, y = k(u^2 - v^2), w = k(u^2 + v^2),$$

where $k \in N, u > v, gcd(u, v) = 1$, and u, v are of a opposite parity.

In the present case we require

$$w = 2z = 2k(u^2 + v^2),$$

so that w is even. This condition holds provided that k is an even positive integer. By substitution, it follows immediately that every triple (x, y, w) obtained from these formulas yields a valid solution of Equation (3.27), where www is an even integer.

For illustration, choosing small parameters gives the following examples:

For $u = 2, v = 1, k = 2$: $x = 8, y = 6, w = 10, z = 5$

For $u = 3, v = 2, k = 2$: $x = 24, y = 10, w = 26, z = 13$.
For $u = 4, v = 3, k = 2$: $x = 48, y = 14, w = 50, z = 25$

and so on.

Thus, all integer solutions to Equation (3.27) arise from the classical parameterization of Pythagorean triples, subject to the additional restriction that the scaling factor k must be even, ensuring that $w = 2z$ is even.

Diophantine equation $x^2 + y^2 = \alpha z^2$

Case $\alpha = 5$

We consider the Diophantine equation

$$x^2 + y^2 = 5z^2. \tag{3.28}$$

Solution.

A trivial solution is $x = \pm 2k, y = \pm k, z = \pm k, k \in Z$.

To obtain further solutions, let us parametrize

$x = 2k + l, y = k - n, z = k$, where $k, l, n \in Z$.

Substituting into (3.28) yields

$$(2k + l)^2 + (k - n)^2 = 5k^2. \tag{3.29}$$

Expanding gives

$$4k^2 + k^2 + l^2 + n^2 + 2k(2l - n) = 5k^2,$$

or equivalently,

$$n^2 - 2kn + l^2 + 4kl = 0.$$

This is a quadratic in n with solutions

$$n_{1,2} = k \pm \sqrt{(k^2 - l^2 - 4kl)},$$

Thus, integrality requires

$$k^2 - l^2 - 4kl = a^2, a \in Z. \tag{3.30}$$

Rearranging, we obtain a quadratic in l:

$$l^2 + 4kl + a^2 - k^2 = 0. \tag{3.31}$$

Its solutions are

$$l_{1,2} = -2k \pm \sqrt{(5k^2 - a^2)}.$$

Hence the discriminant $D = 5k^2 - a^2$ must be square, say $D = b^2$ with $b \in Z$.

Step 1. Let $k = \omega a$ for some $\omega \in Z$. Then

$$D = a^2(5\omega^2 - 1). \tag{3.32}$$

Thus $5\omega^2 - 1$ must be perfect square.

Table 6 lists values of $5\omega^2 - 1$. depending on ω; see Table 6.

Table 6. Table for solution of equation (3.28)

ω	1	2	3	4	5	6	7	8
$\mathbf{5\omega^2 - 1}$	4	19	44	79	124	149	244	319
ω	9	10	11	12	13	14	15	16
$\mathbf{5\omega^2 - 1}$	404	499	604	719	844	979	1124	1279

Ω	17	18	19	20	21	22
$\mathbf{5\omega^2 - 1}$	1444	1619	1804	1999	2204	2419

Inspection shows that only $\omega = 1$ and $\omega = 17$ yield squares ($2^2$and 38^2, respectively).

- For $\omega = 1$, we recover the trivial solution

$(x, y, z) = (\pm 2k, \pm k, \pm k)$.

- For $\omega = 17$, we obtain the nontrivial family

$(x, y, z) = (\pm 38\varepsilon, \pm\varepsilon, 17\varepsilon), \varepsilon \in Z$.

Step 2.

Suppose k is a rational multiple of a, with

$\omega = \frac{2r-1}{2}, r \in N$.

Then

$D = a^2(5\omega^2 - 1)$,

For integrality, $5\omega^2 - 1$ must be the square of a rational number.

Table 7 represents the calculation of the expression $5\omega^2 - 1$ depending on ω; see Table 7.

Table 7. Below presents values of $\mathbf{5\omega^2 - 1}$ for selected values of $\boldsymbol{\omega}$

ω	$\frac{1}{2}$	$\frac{3}{2}$	$\frac{5}{2}$	$\frac{7}{2}$	$\frac{9}{2}$	$\frac{11}{2}$	$\frac{13}{2}$	$\frac{15}{2}$	$\frac{17}{2}$	$\frac{19}{2}$	$\frac{21}{2}$	$\frac{23}{2}$
$5\omega^2 - 1$	$\frac{1}{4}$	$\frac{41}{4}$	$\frac{121}{4}$	$\frac{241}{4}$	$\frac{401}{4}$	$\frac{601}{4}$	$\frac{841}{4}$	$\frac{1121}{4}$	$\frac{1441}{4}$	$\frac{1801}{4}$	$\frac{2201}{4}$	$\frac{2641}{4}$

Table 7 lists corresponding values. It is found that only $\omega = \frac{1}{2}$ and $\omega = \frac{13}{2}$ yield rational squares. No other values produce valid solutions.

From the Table 3.7 we observe that $5\omega^2 - 1$ is the square of the rational numbers (entries in bold) for $\omega = \frac{1}{2}$ and $\omega = \frac{13}{2}$.

Direct calculation, no other value of ω from the given set yields a square of a rational number.

Case 1: $\omega = \frac{1}{2}$.

Then $k = \frac{1}{2}a$. We compute:

$$l_{1,2} = -2k \pm \sqrt{(4k^2 + k^2 - a^2)} = -2k \pm \sqrt{(5k^2 - a^2)}$$

This gives:

$$l_1 = -\frac{a}{2}, l_2 = -\frac{3}{2}a \Rightarrow \omega = \frac{1}{2}\ n_1 = \frac{3}{2}a, n_2 = -\frac{a}{2}.$$

Let be $\varepsilon = \frac{a}{2}$ so $k = \varepsilon$. Substituting into equation (3.29):

$$(2\varepsilon - \varepsilon)^2 + (\varepsilon - 2\varepsilon)^2 = 5\varepsilon^2$$

or:

$$(2\varepsilon - 3\varepsilon)^2 + (\varepsilon + \varepsilon)^2 = 5\varepsilon^2 \tag{3.33}$$

This yields the solution:

$$x = \pm\varepsilon, y = \pm 2\varepsilon, z = \pm\varepsilon, \varepsilon \in Z,$$

which is a known trivial equation.

Case 2: $\omega = \frac{5}{2}$.

Then $k = \frac{5}{2}a = 5\varepsilon$. Substituting into the discriminant expression (the part under the square root) we get

$$l_{1,2} = -2.5\varepsilon \pm \sqrt{(5(5\varepsilon)^2 - 4\varepsilon^2)} = \varepsilon(-10 \pm 11), \tag{3.34}$$

yields:

$$l_1 = \varepsilon, l_2 = -21\varepsilon.$$

Now we compute:

$$n_{1,2} = 2\omega\varepsilon \pm \sqrt{(\omega^2 4\varepsilon^2 - \varepsilon^2 - 4.2\omega\varepsilon\varepsilon)} =$$
$$5\varepsilon \pm \varepsilon\sqrt{\left(\frac{25}{4}.4 - 1 - 8.\frac{5}{2}\right)}$$

$$= 5\varepsilon \pm \varepsilon\sqrt{25 - 1 - 20} = \varepsilon(5 \pm 2),$$

so that;

$n_1 = 7\varepsilon$ and $n_2 = 3\varepsilon$.

Substituting into equation (3.29):

$$(\pm 11\varepsilon)^2 + (\pm 2\varepsilon)^2 = 5(5\varepsilon)^2. \tag{3.35}$$

Thus, another solution is:

$x = \pm 11\varepsilon\,, y = \pm 2\varepsilon, z = 5\varepsilon, \varepsilon \in Z.$

Case 3: $\omega = \frac{13}{2}$

Then $k = \frac{13}{2}a = 13\varepsilon$ and substituting into the discriminant expression:

$$l_{1,2} = -2.13\varepsilon \pm \sqrt{\left(4(13\varepsilon)^2 + (13\varepsilon)^2 - \frac{(2\varepsilon)^2}{169}\right)} = \varepsilon(-26 \pm 29),$$

which gives:

$l_1 = 3\varepsilon\,, l_2 = -55\varepsilon.$

We now compute:

$$n_{1,2} = \omega a \pm \sqrt{(\omega^2 4\varepsilon^2 - 9\varepsilon^2 - 4.2\omega\varepsilon 3\varepsilon)}.$$

Substituting $\omega = \frac{13}{2}, a = 2\varepsilon$, and simplifying, we obtain:

$$13\varepsilon \pm 2\varepsilon\sqrt{\left(\frac{169-9-4.13.3}{4}\right)}=13\varepsilon \pm 2\varepsilon.$$

Therefore:

$n_1 = 15\varepsilon$ and $n_2 = 11\varepsilon$.

Now we substitute l_1, n_1, l_2, n_2 into equation (3.29) and obtain

$$(\pm 29\varepsilon)^2 + (\pm 2\varepsilon)^2 = 5(13\varepsilon)^2. \quad (3.36)$$

Thus, the solution is:

$x = \pm 29\varepsilon\,, y = \pm 2\varepsilon, z = \pm 13\varepsilon, \varepsilon \in Z.$

Solution 2

Obviously, if (x, y, z) is an integer solution of the given equation, then the number $x^2 + y^2$ must be divisible by the number 5. Since every integer c can be expressed in just one of the forms

$c = 5q, 5q + 1, 5q + 2, 5q + 3, 5q + 4$ where $q \in Z$,

we compute the square:

$c^2 = 25q^2, 25q^2 + 10.q + 1, 25q^2 + 20q + 4,25.q^2 + 30\text{q} + 9, 25.q^2 + 40.q + 16.$

Thus, modulo 5, the square of any integer belongs to the remainder class $0, 1,$ or 4. Consequently, the sum $x^2 + y^2$modulo 5 can attain only certain values, as shown in the following table; see Table 8.

Table 8. Remainder classes modulo 5 for $\boldsymbol{x^2 + y^2}$

	Remainder class					
x^2	0	0	0	1	1	4
y^2	0	1	4	1	4	4
$x^2 + y^2$	0	1	4	2	0	3

From the table, only the first column $(0, 0, 0)$ and the fifth column $(1, 4, 0)$ yield viable options for $x^2 + y^2 \equiv 0\ mod\ 5$. The first column leads to the trivial solution:

$x = \pm 2k, y = \pm k, z = \pm k, k \in Z.$

The fifth column offers space for further exploration. It leads to the following types of equations:

$$(5x_1+1)^2+(5y_1+2)^2=5z^2 \quad (3.37)$$

$$(5x_1+1)^2+(5y_1+3)^2=5z^2 \quad (3.38)$$

$$(5x_1+4)^2+(5y_1+2)^2=5z^2 \quad (3.39)$$

$$(5x_1+4)^2+(5y_1+3)^2=5z^2 \quad (3.40)$$

The corresponding solutions are tabulated below. Only rows where $x^2+y^2=5z^2$ with integer z are highlighted in **bold**; see Tables 9, 10 and 11.

Table 9. Integer solutions for equation (3.37)

x_1	y_1	x	y	$x^2+y^2=5.z^2$	z
0	0	1	2	1+4=5.1	1
1	0	6	2	36+4=5.8	
0	1	1	7	1+49=5.10	
1	1	6	7	36+49=5.13	
2	0	11	2	121+4=5.$\mathbf{5^2}$	5
2	1	11	7	121+49=5.34	
1	2	6	12	36+144=5.$\mathbf{6^2}$	6
2	2	11	12	121+144=5.23	

Table 10. Integer solutions for equation (3.38)

x_1	y_1	x	y	$x^2+y^2=5.z^2$	z
0	0	1	3	1+9=5.2	
1	0	6	3	36+9=5.$\mathbf{3^2}$	3
0	1	1	8	1+64=5.13	
1	1	6	8	36+64=5.20	
2	0	11	3	121+9=5.26	
2	1	11	8	121+64=5.37	
1	2	6	13	36+169=5.41	
3	0	16	3	256+9=5.53	
3	1	16	8	256+64=5.$\mathbf{8^2}$	8
3	2	16	13	256+169=5.85	
0	3	1	18	1+324=5.65	

Table 11. Integer solutions for equation (3.39)

x_1	y_1	x	y	$x^2 + y^2 = 5.z^2$	z
0	0	4	2	16+4=5.$\mathbf{2^2}$	2
1	0	9	2	36+4=5.8	
0	1	4	7	16+49=5.13	
1	1	9	7	36+64=5.20	
2	0	14	2	121+9=5.26	
2	1	14	7	121+64=5.37	
1	2	9	12	36+169=5.41	
2	2	14	12	196+144=5.68	
3	0	19	2	361+4=5.74	
3	1	19	7	361+49=5.82	
3	2	19	12	361+169=5.	
0	3	4	17	16+289=5.106	
1	3	9	17	81+289=5.74	
2	3	14	17	196+289=5.97	
3	3	19	17	361+289=5.130	
4	0	24	2	576+4=5.116	
4	1	24	7	576+49=5.125	
4	2	24	12	576+144=5.$\mathbf{12^2}$	12
4	3	24	17	576+289=5.173	
0	4	4	22	16+484=5.$\mathbf{10^2}$	10
1	4	9	22	81+484=5.113	
2	4	14	22	196+484=5.136	
3	4	19	22	361+484=5.$\mathbf{13^2}$	13
4	4	24	22	576+484=5.212	
5	0	29	2	841+4= $\mathbf{5.13^2}$	13
5	1	29	7	841+49=5.178	
5	2	29	12	841+144=5.197	
5	3	29	17	841+289=5.226	
5	4	29	22	841+484=5.1325	
0	5	4	27	16+729=5.149	
1	5	9	27	81+729=5.142	
2	5	14	27	196+729=5.185	
3	5	19	27	361+729=5.218	
4	5	24	27	576+729=5.261	
5	5	29	27	841+729=5.314	
6	0	34	2	1156+4=5.232	
6	1	34	7	1156+49=5.221	
6	2	34	12	1156+144=5.260	
6	**3**	**34**	**17**	**1156+289=5.** $\mathbf{17^2}$	**17**
6	4	34	22	1156+484=5.328	
6	5	34	27	1156+729=5.377	
0	6	4	32	16+1024=5.208	
1	6	9	32	81+1024=5.221	
2	6	14	32	196+1024=5.244	

x_1	y_1	x	y	$x^2 + y^2 = 5.z^2$	z
3	6	19	32	361+1024=5.277	
4	6	24	32	576+1024=5.320	
5	6	29	32	841+1024=5.373	
6	6	34	32	1156+1024=5.436	

Only the highlighted rows correspond to exact solutions. Other rows were omitted as they do not yield integer solutions for z.

3.4.4. Summary of Additional Solutions

We thus obtain additional solutions of the form:

$$(x, y, z) = (k, k, k)\text{and } (x, y, z) = (7k, k, 5k),\text{where } k \in Z.$$

These results match the solutions derived in Solution 1, confirming their validity.

Rows marked bold represent the search solutions.

So, we have follow-up solutions $(x, y, z) = (k, k, k)$ and $(x, y, z) = (7k, k, 5k)$, where $k \in Z$, which are the same solutions as the previous ones obtained in the solution 1.

Remark. We have just listed part of the table, because there are no integer solutions for x, y, z in the other rows.

Conclusion

In this section, we have studied special Diophantine equations of the second degree of the form

$$x^2 + y^2 = \alpha z^2$$

with respect to their solvability.

We began with the classical case $\alpha = 1$, which reduces to the well-known Pythagorean equation—often referred to as the "queen" of Diophantine equations—due to its foundational role in number theory and geometry.

For $\alpha = 0$, the equation has only the trivial solution. When $\alpha < 0$, the equation also admits only trivial integer solutions.

In the subsequent analysis, we explored values $\alpha = 2, 3, 4, 5$ and derived specific solution patterns and conditions for solvability. Results were summarized in detailed tables and supported by illustrative examples.

As a natural continuation, we propose extending this investigation to further values $\alpha = 6, 7, 8, 9$, etc., to uncover additional patterns and generalizations in the behavior of these equations.

References

[1] Choudhry, A. Quadratic Diophantine equations, with applications to quartic equations. *Mountain Journal of Mathematics, 46*(3), 469–804. Retrieved from https://projecteuclid.org (2016).

[2] Dorin, A., & Titu, A. *Quadratic Diophantine Equations.* (Developments in Mathematics, Vol. 40). Cham: Springer. https://doi.org/10.1007/978-3-319-23045-1 (2015).

[3] Medvid', V. Methods of solving some Diophantine equations of the second degree. In *ICERI 2021 Proceedings: 14th Annual International Conference of Education, Research and Innovation* (pp. 3059–3068). Valencia: IATED. ISBN 978-84-09-34549-6. ISSN 2340-1095. (2021).

Chapter 4

Linear Diophantine Equations about n Unknowns

Abstract

Diophantine equations occupy an important place in algebra and number theory. They continue to inspire students at all levels of mathematics education, from elementary school to university. The aim of this chapter is to present methods for solving linear Diophantine equations in nnn unknowns, particularly with an educational focus.
Two main approaches will be discussed:

- the method of solving congruences,
- the method of reduction to fewer unknowns.

Both methods are explained in detail and illustrated with worked examples. At the end of the chapter, several unsolved exercises are provided for independent practice.
Method of solving congruences:
When calculating specific examples, we will proceed similarly to the previous proof, that is, we will solve the congruence according to the modulus, which is the greatest common divisor of $n - 1$ coefficients, provided that it is different from 1, then we will solve the congruence with one unknown, and subsequently, we will gradually calculate the remaining variables.
Method of reduction to a smaller number of unknowns:
The following procedure was proposed by the supervisor. The advantage of this method lies in the fact that we will not need congruences; we will only need the extended Euclidean algorithm.
Educational goal. These methods were presented at selection seminars to talented high school students and university students studying mathematics, and then tests were conducted to evaluate their ability to solve specific tasks related to the mentioned issue.
The students perfectly managed almost all assigned tasks with excellent results.
The aforementioned methods for solving the tasks were also studied with interest by secondary school and university teachers.

With respect to the educational goals, this chapter is recommended for students in mathematical classes and students of mathematics with a specialization in algebra or number theory.

Keywords: linear Diophantine equation, greatest common divisor, congruence, reduction, positive integer, extended Euclidean algorithm

4.1. Introduction

Definition 4.1. A linear Diophantine equation in n unknowns is an equation of the form

$$a_1x_1 + a_2x_2 + \cdots + a_nx_n = b, \tag{4.1}$$

where $a_1, a_2, \ldots, a_n, b$ are integers, and $a_i \neq 0$ for all $i = 1, 2, \ldots, n$ and $x_1, x_2, \ldots x_n$ are unknown, n is a positive integer.

Denote $\delta = gcd(a_1, a_2, \ldots, a_n)$, then we can write $a_i = \delta \,.\, a_i', a_i'$ is an integer, for all $i = 1, 2, \ldots, n$. Then the number δ in equation (4.1) can be pointed out, and we get $\delta.(a_1'x_1 + a_2'x_2 + \cdots + a_n'x_n) = b$ from which we see that for any integers $x_1, x_2, \ldots x_n$ the left side of the equation is divisible by the number δ, in order for this equation to have an integer solution, $\delta | b$.

Then we can write

$$a_1'x_1 + a_2'x_2 + \cdots + a_n'x_n = b' \tag{4.2}$$

where $b' = \delta \,.\, b$ and, moreover, $gcd(a_1', a_2', \ldots, a_n') = 1$. Therefore, if equation (4.1) has an integer solution, it can always be converted to equation (4.2). See e.g Amer [1], Bradler [2], Imomov [3], Mestechkin [4], Medvid [5] or Sanched-Roselly Navaro [6].

Definition 4.2. If in equation (4.1) $gcd(a_1, a_2, \ldots, a_n) = 1$, then we will say that this equation is in basic form.

Theorem 4.1. Let n be a positive integer, $n \geq 2$. Equation (4.1) has the basic form always an integer solution.

Proof.

We perform it by induction with respect to n. So we have to prove the following:

(I) We can prove the statement for $n = 2$.

(II) We will assume that the theorem holds for $n - 1$ and show that it also holds for n.

ad(I) We solve the equation $a_1x_1 + a_2x_2 = b$, which is a linear Diophantine equation that we devoted to the previous chapter, from which, according to Theorem 3.1, we know that this equation will always have an integer solution, since we have $gcd(a_1, a_2) = 1$.

ad(II) Now let $n > 2$. We assume that the theorem holds for equations with $n - 1$ unknowns and prove it for equation (4.1) with n unknowns.

Let us denote $\delta = gcd(a_1, a_2, . . . , a_{n-1})$. Let the integers $x_1, x_2, ... x_n$ are solutions, then they must satisfy the congruence

$$a_1x_1 + a_2x_2 + \cdots + a_nx_n \equiv b \ (mod\ \delta).$$

Because $\delta | a_i$, for all i = 1, 2, . . . , n - 1 we get

$$a_nx_n \equiv b \ (mod\ \delta) \tag{4.3}$$

Since $gcd(\delta, a_n) = gcd(a_1, a_2, . . . , a_n) = 1$, the congruence (4.3) has a solution according to Theorem 3.13, and we can adjust it to the form $x_n \equiv c \ (mod\ \delta)$, where c is some integer, or $x_n = c + \delta t$, where t is an integer. By substituting into equation (4.1), we obtain

$$a_1x_1 + a_2x_2 + \cdots + a_n(c + \delta t) = b$$

$$a_1x_1 + a_2x_2 + \cdots + a_{n-1}x_{n-1} = b - a_nc - a_n\delta t$$

$$\delta(a_1x_1 + a_2x_2 + \cdots + a_{n-1}x_{n-1}) = b - a_nc - a_n\delta t$$

We can divide the equation by the number. δ because it is $a_n c \equiv b \ (mod\ \delta)$, i.e., $\delta |\ (b - a_n c)$. We thus obtain $a'_1 x_1 + a'_2 x_2 + \cdots + a'_{n-1} x_{n-1} = b'$, where $b' = \frac{b - a_n c}{d} - a_n t$.

In this equation, $gcd(a'_1, a'_2, \ldots, a'_{n-1}) = 1$ and, according to assumptions, this equation has an integer solution.

We will only pay attention to the Diophantine equations about n unknowns for $n \geq 3$, the case of $n = 2$ was dealt with in detail in the previous chapter, and we will also consider equation (4.1) such that $a_i \neq 1$ for all $i = 1, 2, \ldots, n$.

If for some $k \in \{1, 2, \ldots, n\}$ $a_k = 1$, then the solution of this equation is a number of the form

$$
\begin{aligned}
&x_1 = t_1 \\
&x_2 = t_2 \\
&\ldots \\
&x_k = b - a_1 t_1 - a_2 t_2 - \cdots - a_n t_n \\
&\ldots \\
&x_n = t_n,
\end{aligned}
$$

where $t_1, t_2, \ldots, t_{k-1}, t_{k+1}, \ldots, t_n$ are integers.

4.2. The Method of Solving Congruences

When calculating specific examples, we will proceed similarly to the previous proof, that is, we will solve the congruence according to the modulus, which is the greatest common divisor of $n - 1$ coefficients, provided that it is different from 1, then we will solve the congruence with one unknown and subsequently we gradually calculate the remaining variables.

For the clarity of the calculation, in the case of a small number of unknowns, we will use the notation for the unknowns $x, y, z, u,$ etc. instead of$x_1, x_2, \ldots x_n$.

Example 4.1. $12x - 9y + 5z = 13$

Any integer solution must necessarily satisfy the following congruence, which we will immediately start modifying

$12x - 9y + 5z \equiv 13\ (mod\ 3)$
$2z \equiv 4\ (mod\ 3)$
$z \equiv 2\ (mod\ 3)$
$z = 2 + 3t,$

where t is an integer. We substitute the solution for z into the given equation.

$12x - 9y + 5 \cdot (2 + 3t) = 13$

$-9y + 10 + 15t = 13$
$12x - 9y = 3 - 15t$
$4x - 3y = 1 - 5t$ (4.4)

This gave us a linear Diophantine equation with two unknowns that we can solve from the previous chapter.

$4x - 3y \equiv 1 - 5t\ (mod\ 4)$
$y \equiv 1 + 3t\ (mod\ 4)$
$y = 1 + 3t + 4s, s \in Z.$

By substituting for y in equation (4.4), we get an equation for x only, which we solve

$4x - 3(1 + 3t + 4s) = 1 - 5t$
$4x - 3 - 9t - 12s = 1 - 5t$
$4x = 4 + 4t + 12s$
$x = 1 + t + 3s$

The solution is, therefore, all numbers x, y, z that are of the form

$x = 1 + t + 3s$
$y = 1 + 3t$
$z = 2 + 3t, t, s \in Z.$

Example 4.2. $17x + 5y - 3z = 4$

This example differs from the previous one in that, for any pair of coefficients, the greatest common divisor of these coefficients is equal to one.

Therefore, we will proceed by choosing one of the coefficients as a module, thereby eliminating one unknown. Then, we will choose the remaining unknowns as integer parameters, using these to complete the example.

$$17x + 5y - 3z \equiv 4\ (mod\ 3)$$
$$2x + 2y \equiv 4\ (mod\ 3)$$
$$x + y \equiv 2\ (mod\ 3)$$

Let $y = t$, where t is an integer, then

$$x + t \equiv 2\ (mod\ 3)$$
$$x \equiv 2 - t\ (mod\ 3)$$
$$x = 2 - t + 3s,$$

where s is an integer. By substituting into the assignment, we get the solution for z.

$$17 \cdot (2 - t + 3s) + 5t - 3z = 4$$
$$3z = 34 - 17t + 51s + 5t - 4$$
$$3z = 30 - 12t + 51s$$
$$z = 10 - 4t + 17s$$

In total

$$x = 2 - t + 3$$
$$= t$$
$$z = 10 - 4t + 17s, t, s \in Z.$$

Example 4.3. $-14x - 2y + 5z + 3u = 2$

We will proceed similarly to the previous example.

$$-14x - 2y + 5z + 3u \equiv 2\ (mod\ 5)$$
$$x - 2y + 3u \equiv 2\ (mod\ 5)$$

Let $y = t$ and $u = s$, where t, s are integers, then

$$x - 2t + 3s \equiv 2\ (mod\ 5)$$

$$x \equiv 2 + 2t - 3s \pmod 5$$
$$x = 2 + 2t - 3s + 5r,$$

where r is an integer. By substituting, we calculate the last unknown.

$$-14 \cdot (2 + 2t - 3s + 5r) - 2t + 5z + 3s = 2$$
$$-28 - 28t + 42s - 70r - 2t + 5z + 3s = 2$$
$$5z = 30 + 30t - 45s + 70y$$
$$z = 6 + 6t - 9s + 14y$$

Together

$$x = 2 + 2t - 3s + 5y$$
$$y = t$$
$$z = 6 + 6t - 9s + 14y$$
$$u = s, t, s, r \in Z.$$

4.3. The Method of Reduction to a Smaller Number of Unknowns

The following procedure was proposed by the supervisor. The advantage of this method lies in the fact that we will not need congruences; we will only need the extended Euclidean algorithm.

We limit ourselves to the case when equation (4.1) is in the basic form, because if it is not and has an integer solution, we can convert it to this form. I remind you that we consider $n \geq 3$.

We define $\delta = gcd(a_1, a_2, \cdots, a_{n-1})$ and put $a_i' = \frac{a_i}{\delta}$ for $i = 1, 2, \ldots, n-1$.

$$a_1x_1 + a_2x_2 + \cdots + a_{n-1}x_{n-1} + a_nx_n = b$$
$$a_1'\delta x_1 + a_2'\delta x_2 + \cdots + a_{n-1}'\delta x_{n-1} + a_nx_n = b$$
$$\delta(a_1'x_1 + a_2'x_2 + \cdots + a_{n-1}'x_{n-1}) + a_nx_n = b$$

If we make the substitution $y = a_1'x_1 + a_2'x_2 + \cdots + a_{n-1}'x_{n-1}$, we obtain a linear Diophantine equation with two unknowns

$$\delta y + a_nx_n = b,$$

thereby converting the problem into a case that we can solve. After finding the solutions for y, x_n we will solve the equation

$$a_1'x_1 + a_2'x_2 + \cdots + a_{n-1}'x_{n-1} = y,$$

where we have one less unknown than at the beginning, and will have an integer solution, since $gcd(a_1, a_2, \cdots, a_{n-1}) = 1$. We can repeat the entire procedure until equation (4.1) we do not reduce to a linear Diophantine equation with two unknowns that we can solve. This gives us all the solutions to equation (4.1).

Let's show the above procedure for $n = 3$. We solve the equation in the basic form

$$ax + by + cz = \delta,$$

where a, b, c, δ are integers, $a > 1, b > 1, c > 1$.

Denote $D = gcd(a, b), then\ a = Da', b = Db'$,

where a', b' are positive integers and $gcd(a', b') = 1, gcd(D, c) = 1$.

$$ax + by + cz = \delta$$
$$Da'x + Db'y + cz = \delta$$
$$D(a'x + b'y) + cz = \delta$$
$$Du + cz = \delta$$

We denote $a'x + b'y = u$.

Using the extended Euclidean algorithm, we find the numbers u_0, z_0 satisfying $Du_0 + cz_0 = 1$. Then $u = \delta u_0 - pc, z = dz_0 + pD$, where p is an integer. Now we solve the equation $a'x + b'y = \delta u_0 - pc$. Again, using the extended Euclidean algorithm, we find integers x_0, y_0 satisfying $a'x_0 + b'y_0 = 1$.

Then $x = x_0(\delta u_0 - pc) - b'q$, $y = y_0(\delta u_0 - pc) - a'q$, where q is an integer.

The integers x, y, z solve the equation $ax + by + cz = \delta$ if and only if

$$x = x_0\delta u_0 - x_0 pc - b'q$$

$$y = y_0\delta u_0 - y_0\, pc - a'q$$
$$z = z_0 d + pD, p, q \in Z.$$

Example 4.4. $12x + 9y + 5z = 4$

It is $gcd(12, 9) = 3$, then

$$12x + 9y + 5z = 4$$
$$3(4x + 3y) + 5z = 4$$
$$3u + 5z = 4,$$

where we introduced the substitution $4x + 3y = u$. We solve the equation $3u + 5z = 4$.

Using the extended Euclidean algorithm, we obtain the following Bezout equality, which we will modify further

$$3(2) + 5(-1) = 1$$
$$3(8) + 5(-4) = 4,$$

the solution of the equation $3u + 5z = 4$ is then the number of the form

$$u = 8 + 5p\; z = -4 - 3p,$$

where p is an integer. Returning to substitution, we must solve the equation

$$4x + 3y = 8 + 5p.$$

Again, using the extended Euclidean algorithm, we obtain Bezout's equality

$$4(1) + 3(-1) = 1$$
$$4(8 + 5p) + 3(-8 - 5p) = 8 + 5p,$$

then the shape numbers are the solution

$$x = 8 + 5p - 3q\; y = -8 - 5p + 4q,$$

where q is an integer. The solutions to the equation $12x + 9y + 5z = 4$ are integers of the form

$$x = 8 + 5p - 3q$$
$$y = -8 - 5p + 4q$$
$$z = -4 - 3p, p, q \in Z.$$

Exercise 4.1.

Solve the Diophantine equations:

$(a)\ 12x + 16y - 7z = 2$
$(b)\ 8x - 32y + 48z - 3u = 12$
$(c)\ 12x + 13y - 4z + 2u = 25$
$(d)\ 5x + 15y - 25z = 4$

References

[1] Amer, R. *The Mathematical Analysis of Linear Diophantine Equations*. [Journal article]. (2023).

[2] Bradler, K. (n.d.). On the Number of Nonnegative Solutions of a System of Linear Diophantine Equations. *arXiv* preprint. https://arxiv.org.

[3] Imomov, A. On Some Methods for Solution of Linear Diophantine Equations. *DergiPark*. https://dergipark.org.tr (2016).

[4] Mestechkin, M. M. Natural solutions of Diophantine linear equation with n unknowns. *Journal of Computational Mathematics*. ACM Digital Library. (2016).

[5] Medviď, V. Linear Diophantine Equations about n Unknowns. In *Proceedings of the International Conference on Education and New Learning Technologies*. IATED Digital Library. https://library.iated.org (2024).

[6] Sánchez-Roselly Navarro, A. *Linear Diophantine Equations and Applications.* University of Granada. Retrieved from https://digibug.ugr.es (2016).

Chapter 5

Pythagorean Trinity

Abstract

We call positive integers x, y, z a Pythagorean triplet if they satisfy the equation

$$x^2+y^2=z^2$$

This equation is a special case of equation

$$x^n+y^n=z^n$$

where n is a positive integer. This equation is known as Fermat's equation and belongs to one of the most famous Diophantine equations, due to its interesting history. This equation for $n = 2$ was already addressed in ancient Greece, and it was known to have infinitely many solutions. The French mathematician Pierre Fermat, who lived in the 17th century, believed that there are no positive integers that satisfy equation for $n \geq 3$. This hypothesis is known as Fermat's Big Theorem. Fermat had a note in the margin of one of his books that he had found a proof of this theorem, but it is said to be too long to write here. However, his proof has never been found, but we know that he managed to prove this statement for n = 4. Nevertheless, the final general proof was brought by the British mathematician Andrew Wiles in 1994, which is the longest proof in the history of mathematics.

Educational goal. This chapter is intended for students of universities with a mathematical focus as a perfect tool for further theoretical scientific research in the field of algebra and number theory.

Keywords: Pythagorean equation, algorithm, different parity, coprime number, equation, equality, coefficient, relation, integer solution

5.1. Introduction

We call positive integers x, y, z a Pythagorean triple if they satisfy the equation

$$x^2 + y^2 = z^2 \tag{5.1}$$

Equation (5.1) is a special case of equation

$$x^n + y^n = z^n \tag{5.2}$$

where n is a positive integer. This equation is known as Fermat's equation and belongs to one of the most famous Diophantine equations, due to its fascinating history. This equation for $n = 2$ was already addressed in ancient Greece, and it was known that it would have an infinite number of solutions. The French mathematician Pierre Fermat, who lived in the 17th century, believed that there are no positive integers that satisfy equation (5.2) for $n \geq 3$. This hypothesis is known as Fermat's Big Theorem. Fermat had a note in the margin of one of his books that he had found a proof of this theorem, but it is said to be too long to write here. However, his proof has never been found, but we know that he managed to prove this statement for n = 4. Although the final general proof was brought by the British mathematician Andrew Wiles in 1994, it is the longest proof in the history of mathematics. In the appendix of this text, we will show the non-existence of a solution equation (6.2) for $n = 4$.

We will now solve equation (5.1). We use the known identity

$$1 + 2a + a^2 = (1 + a)^2 \tag{5.3}$$

which we could compare with the investigated equation (5.1). We would need to $1 + 2a$ was the square of some positive integer t, so we require $t^2 = 1 + 2a$, from here, we express $a = \frac{1}{2} \cdot (t^2 - 1)$ and substitute in (5.3).

$$t^2 + \frac{1}{4}(t^2 - 1)^2 = \frac{1}{4}(t^2 + 1)^2$$
$$(2t)^2 + (t^2 - 1)^2 = (t^2 + 1)^2$$

By further comparison, we get

$$x = 2t$$
$$y = t^2 - 1$$
$$z = t^2 + 1,$$

where t is a positive integer, $t \geq 2$, because by choosing $t = 1$ we would get $y = 0$, but we are only looking for positive solutions. We can call these equations the generator of Pythagorean triples, see e.g., Borovich and Shafarevich [1], Dickson [2], Medvid' [3], Mordel [4], Schimmel [5] and Watson [6].

Let's see what the first five triples will look like; we get them by choosing

t = 2, 3, 4, 5, 6.

t	x	y	z
2	3	4	5
3	6	8	10
4	8	15	17
5	10	24	26
6	12	35	37

Although we have found an infinite number of triples satisfying equation (5.1), unfortunately they are not all solutions. We can do this by finding a triple that is a solution to the given equation, but we do not get it by any choice of the parameter t, for example

$$5.2 + 122 = 132.$$

We can see that the generator didn't generate this triple for us, and it can't anymore, because the numbers of the searched triples form increasing sequences, and the values are already high on the selection line $t = 3$. So we will look for a generator that will find us all Pythagorean triples.

Lemma 5.1. Let c be odd, or an even number. Then c^2 is also odd or even.

Proof.

If c is odd, it can be written in the form $c = 2k - 1$, where k is an integer. Then, of course,

$$c^2 = (2k - 1)^2 = 4k^2 - 4k + 1 = 4(k^2 - k) + 1,$$

which is again an odd number.

If c is even, it can be written in the form $c = 2l$, where l is an integer. Then, of course,

$$c^2 = (2l)^2 = 4l^2 ,$$

which is again an even number.

Lemma 5.2. Let a, b be positive integers such that $gcd(a, b) = 1$. If $a \cdot b = n^2$, where n is a positive integer, then $a = u^2$ $b = v^2$, where u, v are positive integers.

Proof.

Let's do the prime factorization of the numbers a, b, n:

$$a = p_1^{\alpha_1} p_2^{\alpha_2} \dots p_k^{\alpha_k},$$

where $p_1 < p_2 < \cdots < p_k$ are prime numbers and $\alpha_1, \alpha_2, \dots, \alpha_k$ are positive integers,

$$b = q_1^{\beta_1} q_2^{\beta_2} \dots q_l^{\beta_l},$$

where $q_1 < q_2 < \cdots < q_l$ are prime numbers and $\beta_1, \beta_2, \dots, \beta_l$ are positive integers,

$$n = r_1^{\gamma_1} r_2^{\gamma_2} \dots r_m^{\gamma_m},$$

where $r_1 < r_2 < \cdots < r_m$ are prime numbers and $\gamma_1, \gamma_2, \dots, \gamma_m$ are positive integers.

Moreover, it is $n^2 = r_1^{2\gamma_1} r_2^{2\gamma_2} \dots r_m^{2\gamma_m}$.

Since $gcd(a, b) = 1$, the primes p_h must be different from the primes q_i, for all $h = 1,2, \dots, k$ and $i = 1,2, \dots, l$.

$$n^2 = a.b$$
$$r_1^{2\gamma_1} r_2^{2\gamma_2} \dots r_m^{2\gamma_m} = p_1^{\alpha_1} p_2^{\alpha_2} \dots p_k^{\alpha_k} q_1^{\beta_1} q_2^{\beta_2} \dots q_l^{\beta_l}$$

Let's relabel the right side of the last equality

$$r_1^{2\gamma_1} r_2^{2\gamma_2} \dots r_m^{2\gamma_m} = s_1^{\delta_1} s_2^{\delta_2} \dots k_{k+l}^{\delta_{k+l}}, \tag{5.4}$$

where $s_1 < s_2 < \cdots < s_{k+l}$ are prime numbers and $\delta_1, \delta_2, \ldots, \delta_{k+l}$ are all positive numbers. From there, $r_j | s_1^{\delta_1} s_2^{\delta_2} \ldots k_{k+l}^{\delta_{k+l}}$ for all $j = 1, 2, \ldots, m$, because r_j is prime, then by Lemma 5.1 there exists $s_e^{\delta_e}$, for some $e = 1, 2, \ldots, k + l$ such that $r_j | s_e^{\delta_e}$. Using again that r_j is prime, we get $r_j | s_e$, it is possible, but only if $r_j = s_e$. This means that there are at least as many prime numbers on the right-hand side of equality (5.4) as on the left, i.e., $k + l \geq m$. It can be proved by a completely similar reasoning that $m \geq k + l$, which together gives us $m = k + l$. Since in (5.4) we have the prime numbers ordered by size, $r_1 =, r_2 = s_2, \ldots, r_m = s_m$.

A consequence of these equalities is the equality of the exponents of prime numbers

$$\delta_j = 2\gamma_j,$$

for all $j = 1, 2, \ldots, m$. Since the exponents δ_j are even numbers, α_j, β_j are also even numbers, therefore a, b are square powers of some positive integers u, v.

Lemma 5.3. Let a, b be positive integers. If $b^2 | a^2$, then $b \,| a$.

Proof.

Of the assumptions, $a^2 = u\, b^2$. If we can prove that $u = v^2$, where v is a positive integer, then

$$a^2 = v^2 b^2$$
$$a^2 = (vb)^2 ,$$

hence $a = v\ b$, since $v\ b$ is a positive integer, and therefore $b \,| a$.

We show that $u = v^2$, where v is a positive integer. Let's perform a prime factorization of the numbers a, b, u:

$$a = r_1^{\gamma_1} r_2^{\gamma_2} \ldots r_m^{\gamma_m}$$

where $r_1 < r_2 < \cdots < r_m$ are prime numbers and $\gamma_1, \gamma_2, \dots, \gamma_m$ are positive integers, $b = p_1^{\alpha_1} p_2^{\alpha_2} \dots p_k^{\alpha_k}$, where $p_1 < p_2 < \cdots < p_k$ are prime numbers and $\alpha_1, \alpha_2, \dots, \alpha_k$ are positive integers, $u = q_1^{\beta_1} q_2^{\beta_2} \dots q_l^{\beta_l}$, where $q_1 < q_2 < \cdots < q_l$ are prime numbers and$\beta_1, \beta_2, \dots, \beta_l$ are positive integers.

Moreover,

$$b^2 = p_1^{2\alpha_1} p_2^{2\alpha_2} \dots p_k^{2\alpha_k}$$
$$a^2 = r_1^{2\gamma_1} r_2^{2\gamma_2} \dots r_m^{2\gamma_m}$$

Let's count:

$$a^2 = u \cdot b^2$$

$r_1^{2\gamma_1} r_2^{2\gamma_2} \dots r_m^{2\gamma_m} = s_1^{\delta_1} s_2^{\delta_2} \dots k_{k+l}^{\delta_{k+l}}$, where $s_1 < s_2 < \cdots < s_{k+l}$ are prime numbers and $\delta_1, \delta_2, \ldots, \delta_{k+l}$ are all positive numbers.

Now, completely analogously, as in the proof of the previous lemma, it would be proved that the exponents δ_j are even numbers, for all $j = 1, 2, \ldots, m$. It follows that u is the square of some positive integer v.

Let δ denote the greatest common divisor of the numbers x, y from equation (5.1). We can then write $x = x_1\delta, y = y_1\delta$, where x_1, y_1 are negative positive integers, and by substituting into the equation, we have

$$\delta^2 x_1^2 + \delta^2 y_1^2 = z^2$$
$$\delta^2 (x_1^2 + y_1^2) = z^2,$$

hence $\delta^2 | z^2$ i.e., according to Lemma 5.3 $\delta | z$, but then $z = z_1\delta$, wherez_1 is a positive integer.

$$\delta^2 (x_1^2 + y_1^2) = \delta^2 z_1^2$$
$$x_1^2 + y_1^2 = z_1^2$$

We got the same equation, but abbreviated, i.e., each equation (5.1) can be converted into a form where the greatest common divisor of the numbers x, y will be equal to 1. Therefore, let us consider from this while equation (5.1), where $gcd(x, y) = 1$. Then, however, at least one of the numbers x, y must be odd. Without loss of generality, we can assume that x is an odd number.

$$x^2 = z^2 - y^2 = (z + y)(z - y) \tag{5.5}$$

Denote $\delta' = gcd(z + y, z - y)$.

Then

$$z + y = a\delta' \; z - y = b\delta', \tag{5.6}$$

where a, b are positive integers and of course $gcd(a, b) = 1$. Substituting into equation (5.5)

$$x^2 = ab\delta'^2. \tag{5.7}$$

We can prove by argument that $gcd(ab, \delta'^2) = 1$.

Suppose that $gcd(ab, \delta'^2) \neq 1$, then there exists a prime number p such that $p| \; ab, p \mid \delta'^2$. Because $p \mid ab$ and $x^2 = ab\delta'^2, p \mid x^2$, since p is prime, then by Lemma 5.1, must $p \mid x$, we also know that x is odd, therefore $p \neq 2$.

Because $p \mid \delta'^2$, must $p \mid \delta'$, and therefore from (5.6) it follows, $p \mid (z + y), p \mid (z - y)$, hence $p \mid 2y$ and since we already know that $p \neq 2$, p must $p \mid y$, of course, this is in contradiction with the coprime of the numbers x, y.

Since in (6.7) $gcd(ab, \delta'^2) = 1$, according to Lemma 5.2, $ab = q^2$, where q is a positive integer. Hence, reapplying Lemma 6.2 is $a = u^2 \; b = v^2$, where u, v are positive integers.

Then

$$x^2 = u^2v^2\delta'^2$$
$$x^2 = (uv\delta')^2$$
$$x = uv\delta'$$

Hence, the numbers u, v, δ' must also be odd, because if at least one of them were even, then the number x would also be even, which contradicts our assumption. Summing equations (5.6), we get

$$2z = a\delta' + b\delta'$$
$$2z = u^2\delta' + v^2\delta'$$
$$z = \frac{u^2 + v^2}{2}\delta'$$

Analogously, by subtracting equations (5.6), we obtain

$$2y = a\delta' - b\delta'$$
$$2y = u^2\delta' - v^2\delta'$$
$$y = \frac{u^2 - v^2}{2}\delta'.$$

From this, it follows that $\delta' = 1$, otherwise the numbers x and y would be commensurable, which again would contradict the assumption of their non-evenness. Let's also note that the difference $u^2 - v^2$ is always a positive number, because from the equations (6.6) it follows that $a > b$ and therefore also $u > v$.

From the inequality of the numbers a, b follows the inequality of the numbers u, v. Equality

$$x = uv, y = u^2 - v^2, z = \frac{u^2 + v^2}{2} \tag{5.8}$$

for positive integers, odd $u, v, u > v$ give us all disjoint triples x, y, z that satisfy equation (5.1). All other solutions are obtained by multiplying the solutions that we obtain from formulas (5.8), any positive integer.

For an idea, let's write several Pythagorean triples obtained by the generator (5.8).

v	u	x	y	z
1	3	3	4	5
1	5	5	12	13
1	7	7	24	25
3	5	15	8	17
3	7	21	20	29

Let's give the generator of Pythagorean triples (5.8) another shape that we like to throw later. Let's put

$$r = \frac{u+v}{2}$$
$$s = \frac{u-v}{2}$$

We calculate u , v

$$2r = u + v$$
$$2s = u - v$$

From there, we get the sum, the difference, and equality

$$u = r + s\, v = r - s,$$

it follows from these equalities that r, s must be unequal, otherwise the inequality of the numbers u, v would be violated, moreover, so that the difference, or the sum of the numbers r, s was an odd number; therefore, one of them must be odd and the other even. By substituting into (5.8), we obtain a different form of generators of Pythagorean triples

$$x = r^2 - s^2, y = 2rs, z = r^2 + s^2, \tag{5.9}$$

where r, s are positive integers, $r > s$, odd and one of them must be odd and one must be even. This also shows that y is even.

References

[1] Borevich, Z. I., & Shafarevich, I. R. *Number Theory.* New York & London: Academic Press. (1966).

[2] Dickson, L. E. *History of the Theory of Numbers.* Washington, DC: Carnegie Institute. (1920).

[3] Medvid', V. Methods of solving some Diophantine equations of the second degree. In *ICERI 2021 Proceedings: 14th Annual International Conference of Education, Research and Innovation* (pp. 3059–3068). Valencia: IATED. ISBN 978-84-09-34549-6. ISSN 2340-1095. (2021).

[4] Mordell, L. J. *Diophantine Equations.* New York: Academic Press. (Series: Pure and Applied Mathematics). (1969).

[5] Schimmel, A. *The Mystery of Numbers.* New York: Oxford University Press. (1993).

[6] Watson, G. L. *Integral Quadratic Forms.* Cambridge: Cambridge University Press. (1960).

Chapter 6

Diophantine Equations of Higher Degrees – Fermat's Equation

Abstract

An important French mathematician (civil profession - lawyer), Pierre Fermat (1601-1665), stated that the equation

$$x^{n} + y^{n} = z^{n}$$

for $n \in N, n \geq 3$ has no solution in the set of natural numbers. He claimed to have proof of this claim, but his proof has not been found. Only its proof was found for n=4. Leonhard Euler proved the statement for n=3 and Peter Gustav Lejeune Dirichlet for n=5. Later, the German mathematician Ernst Kummer tried to prove this statement in general. He believed he had succeeded, but after a while, he became convinced that he was wrong. After him, other mathematicians also tried to prove the so-called "great Fermat's theorem." The assertion of this theorem has only been proven in a number of special cases, and it was not until 1994 that the British mathematician Andrew John Wiles proved it in general. This chapter presents a proof for the equation $x^4+y^4=z^2$
Educational goal. This chapter is intended for students of mathematically oriented universities as well as for legal scholars in mathematical institutes and university mathematicians in mathematics departments.

Keywords: Pythagorean equation, different parity, coprime number, equality, relation, integer solution

6.1. Introduction

An important French mathematician (civil profession - lawyer), Pierre Fermat (1601-1665), stated that the equation

$$x^{n} + y^{n} = z^{n} \tag{6.1}$$

for $n \in N, n \geq 3$ has no solution in the set of natural numbers. He claimed to have proof of this claim, but his proof has not been found. Only its proof was found for $n = 4$. Leonhard Euler proved the statement for $n = 3$ and Peter Gustav Lejeune Dirichlet for $n = 5$. Later, the German mathematician Ernst Kummer tried to prove this statement in general. He believed he had succeeded, but after a while, he became convinced that he was wrong. After him, other mathematicians also tried to prove the so-called "great Fermat's theorem." The assertion of this theorem has only been proven in a number of special cases, and it was not until 1994 that the British mathematician Andrew John Wiles proved it in general. See Aczel [1], Edwards [2], Medviď [3], Schwartz and Weil [4], Singh [5], Ribenboim [6] or Ribenboim [7].

We show this proof for $n = 4$.

Sentence. Fermat's equation

$$x^4 + y^4 = z^4 \tag{6.2}$$

has no solution in the set of integers, for $xyz \neq 0$.

Proof. We can make an even stronger claim, namely that the equation

$$x^4 + y^4 = z^2 \tag{6.3}$$

does not have a solution in the set of integers for which $xyz \neq 0$ holds, from which the assertion of the theorem immediately follows. If the triplet of integers $(x, y, z), xyz \neq 0$ would be a solution to equation (6.3), then the triplet (x, y, z^2) will satisfy equation (6.2).

If equation (6.2) has a solution (x, y, z), where $x, y, z \in Z, xyz \neq 0$, then we can assume that the numbers x, y, z are coprime. If the following would apply: $gcd(x, y) = \delta > 1$, so $x = \delta x_1, y = \delta y_1$, where $d(x_1, y_1) = 1$. If we multiply equation (6.2) by the number $\frac{1}{\delta^4}$, we get:

$$x_1^4 + y_1^4 = \left(\frac{z}{\delta^2}\right)^2 = z_1^2 \tag{6.4}$$

Since $x_1, y_1 \in Z$, so is $z_1 = \frac{z}{\delta^2} \in Z$.

If: $d(y_1, z_1) = k > 1$, then due to (6.3) the number x_1^2 would have to be divisible by the number k, which means that the number x_1 and k, i.e., also the numbers x_1, y_1 could not be coprime. We have thus proved that if there is an integer solution of (6.2) for which $xyz \neq 0$ holds, then this equation also has an integer solution (x_1, y_1, z_1) for which $x_1 y_1 z_1 \neq 0$ holds, and every pair of these numbers is coprime. It follows that it suffices to prove that equation (2) does not have an integer solution (x, y, z) for which $xyz \neq 0$ holds, and every pair of these numbers is mutually coprime.

In the previous subsection, we proved that all solutions of the Pythagorean equation (x, y, z), where $x, y, z \in N$ are coprime after two, can be expressed in the form

$$x = uv, y = \frac{u^2 - v^2}{2}, z = \frac{u^2 + v^2}{2} \tag{6.5}$$

where $u, v \in N$ are any odd coprime numbers, $u > v$.

Let us modify the expression of the numbers from (6.5) by substituting

$$\frac{u+v}{2} = a, \frac{u-v}{2} = b \tag{6.6}$$

where $a, b \in N$ are numbers of different parities, $a > b$. From (6.6) we have:

$$u = a + b, v = a - b \tag{6.7}$$

Equalities (6.6) and (6.7) show that any pair of odd indivisible numbers $u, v \in N, u > v$ corresponds to a pair of coprime numbers $a, b \in N, a > b$ of different parity and vice versa.

Therefore, if we substitute u, v in (6.5) from relations (6.7), we get that all solutions (x, y, z) of the Pythagorean equation, where $x, y, z \in N$ are indivisible by two, x is odd, we can express shaped

$$x = a^2 - b^2, y = 2ab, z = a^2 + b^2 \tag{6.8}$$

where $a, b \in N$ are coprime numbers of different parity, such that $a > b$. It follows from relations (6.7) that the numbers x, y are of different parity.

If the equation (6.2) has a solution (x_0, y_0, z_0), then $(x_0^2)^2 + (y_0^2)^2 = z_0^2$ which means that the triplet of numbers (x_0^2, y_0^2, z_0^2) is a solution to the

Pythagorean equation. But that means there are numbers $a, b \in N, a > b$, coprime and of different parity, such that

$$x_0^2 = a^2 - b^2, y_0^2 = 2ab,\ z_0 = a^2 + b^2. \tag{6.9}$$

We assume that x_0 is odd and y_0 is even. If it were the other way around, just replace x_0, y_0 with each other.

If $c \in N$ is any odd number, that is, $c = 2k + 1$, where $k \in N$, then $c^2 = 4k^2 + 4k + 1 = 4(k^2 + k) + 1$, which means that the square of an odd number always has a remainder of 1 when divided by the number 4. Therefore, it follows from the first equality of relation (6.8) that a is odd and b is even. Otherwise, when dividing by the number 4, the left side of this equality would have a remainder of 1 and the right side would have a remainder of -1. Since a is odd and $d(a, b) = 1$, $d(a, 2b) = 1$. But then the equality $y_0^2 = 2ab$ implies that

$$a = t^2\ 2b = s^2 \tag{6.10}$$

where $t, s \in N$. However, from the equality $x_0^2 = a^2 - b^2$, i.e., $x_0^2 + b^2 = a^2$, it follows that the triplet (x_0, b, a) is a solution to the Pythagorean equation, which means that according to (6.7), apply:

$$x_0 = m^2 - n^2, b = 2mn, a = m^2 + n^2,$$

where $m, n \in N$ are some numbers of different parity. From the second equality in relations (6.9), we get $mn = \frac{b}{2} = \left(\frac{s}{2}\right)^2$, from which, due to the coprime of the numbers m, n, it follows that

$$m = p^2, n = q^2 \tag{6.11}$$

where $p, q \in Z, pq \neq 0$. Since $a = t^2$ and $a = m^2 + n^2$, it follows from relations (6.10) that

$$p^4 + q^4 = t^2 \tag{6.12}$$

However, it follows from the third equation (8): $z_0 - a^2 + b^2 > a^2$ that

$$0 < t = \sqrt{a} < \sqrt[4]{z_0} < z_0, \quad (z_0 > 1) \tag{6.13}$$

If we put $q = x_1, p = y_1\ t = z_1$, we find that if there is a solution (x_0, y_0, z_0) of equation (3), then there must also be another solution (x_1, y_1, z_1) of this equation, as shown by the equality (6.11), while $0 < z_1 < z_0$. However, we can repeat this reasoning any number of times and get a sequence of solutions to equation (6.3): $(x_0, y_0, z_0), (x_1, y_1, z_1)$, $\dots, (x_n, y_n, z_n), \dots$, for which:

$$z_0 > z_1 > \cdots z_n > \cdots$$

However, the natural numbers that are at most equal to $z_0 \in N$ cannot form an infinite descending sequence, because there can be at most z_0 members in that sequence. The assumption that equation (6.3) has at least one solution (x, y, z) where $x, y, z \in N, xyz \neq 0$ thus led us to a dispute, which proves that equation (6.3) does not have a non-zero integer solution. However, we have thus proved, as we already stated at the beginning of the proof, that equation (2) does not have an integer solution.

Note 1. We actually proved that not only equation (6.2), but also the equation $x^{4n} + y^{4n} = z^{2n}$, where $n \in N$ is a given number, does not have an integer non-zero solution.

Note 2. It should be noted that the Pythagorean equation has infinitely many integer solutions. One such solution is, e.g., $x = 3, y = 4, z = 5$. The reader can easily find other positive integer solutions of this equation.

References

[1] Aczel, A. D. (1996). *Fermat's Last Theorem: Unlocking the Secret of an Ancient Mathematical Problem.* New York: Delta.

[2] Edwards, H. M. (1977). *Fermat's Last Theorem: A Genetic Introduction to Algebraic Number Theory.* New York: Springer.

[3] Medvid', V. Methods of solving some Diophantine equations of the second degree. In *ICERI 2021 Proceedings: 14th Annual International Conference of Education, Research and Innovation* (pp. 3059–3068). Valencia: IATED. ISBN 978-84-09-34549-6. ISSN 2340-1095. (2021).

[4] Schwartz, L., & Weil, A. (Various years). Discussions of Fermat's Last Theorem. In *Collected Works.*

[5] Singh, S. *Fermat's Enigma: The Epic Quest to Solve the World's Greatest Mathematical Problem.* New York: Walker & Company. (1997).
[6] Ribenboim, P. *13 Lectures on Fermat's Last Theorem.* New York: Springer. (1979).
[7] Ribenboim, P. *Fermat's Last Theorem for Amateurs.* New York: Springer. (1999).

Chapter 7

Diophantine Linear Equations

Abstract

We have already discussed linear Diophantine equations in the first chapter. In this chapter, we have focused on a higher, more demanding theoretical professional level, in addition to the already mentioned statements, definitions, and algorithms presented in the first chapter, as well as introducing new definitions, theorems, and lemmas with proofs and algorithms. All new theoretical knowledge is explained with examples and exercises along with their respective solutions.
This chapter is intended for students of mathematics at universities and also for scientists in the field of algebra and number theory.
Our educational work at universities has demonstrated that students have mastered the given issue perfectly and have successfully solved even the most challenging tasks.
In addition to definitions, theorems, lemmas, linear congruences, and eliminations, the following algorithms are also presented in this chapter:
Algorithm for finding all integer solutions of the equation $ax + by = c$.
Row Elimination and Euclid's Algorithm.
Algorithm for easy finding of all integer solutions of the equation $ax + by = c$.
The last part of this chapter is devoted to special equations in Z_n spaces.

Keywords: linear Diophantine equation, greatest common divisor, least common multiple, congruence, reduction, elimination, positive integer, Euclidean algorithm, Z_n space

7.1. Introduction

The central object of Diophantine analysis is the search for integer solutions to polynomial equations. Among these, linear Diophantine equations represent the most elementary and historically earliest class of such problems. A *linear Diophantine equation* in n variables is an equation of the form

$$a_1x_1 + a_2x_2 + \cdots + a_nx_n = b,$$

where $a_1, a_2, \ldots, a_n, b \in Z$ are given integers and the unknowns $x_1, x_2, \ldots, x_n$ are required to take integer values.

7.1.1. Fundamental Solvability Condition

A classical theorem asserts that the equation is solvable in integers if and only if the greatest common divisor of the coefficients divides the constant term:

$$gc\,d(a_1, a_2, \ldots, a_n) \mid b.$$

This criterion provides both a necessary and sufficient condition for solvability, reflecting the basic principle of divisibility in number theory.

7.1.2. Structure of the Solution Set

When solutions exist, they do not typically form a finite set but rather an infinite family parametrized by integer parameters. For the simplest case of two variables,

$$ax + by = c,$$

if $d = gcd\,(a, b)$, then one particular solution (x_0, y_0) can be obtained (for instance, via the extended Euclidean algorithm). The general solution is then

$$x = x_0 + \frac{b}{d}, y = y_0 - \frac{a}{d}t, \mathrm{t} \in \mathrm{Z},$$

which describes a one-dimensional lattice of integer points. For higher dimensions, the set of solutions forms an affine sublattice of Z_n.

7.1.3. Methods of Solution

The principal methods for solving linear Diophantine equations include:

- Euclidean algorithm: provides explicit solutions in two variables.
- Reduction to particular solution plus homogeneous part: separates the equation into one explicit solution and the set of solutions to the associated homogeneous equation.
- Lattice methods and integer linear algebra: in several variables, solutions are naturally described using the theory of integer lattices, Smith normal form of integer matrices, and module theory over Z.

7.1.4. Interplay with Congruences

Linear Diophantine equations are closely related to linear congruences. For example, the equation

$$ax + by = c$$

may be rewritten modulo b to obtain a congruence

$$ax \equiv c(mod\ b),$$

whose solvability conditions and solutions directly yield integer solutions to the original equation. This correspondence illustrates the deep unity of Diophantine equations and modular arithmetic. See e.g., Chionglo [1], Jones [2] and Verdal [8].

7.1.5. Historical and Theoretical Significance

The study of linear Diophantine equations dates back to antiquity and was systematically developed by Diophantus of Alexandria, after whom such problems are named. Later, the systematic use of the Euclidean algorithm and the work of Fermat, Euler, and Gauss established these equations as fundamental examples in number theory. In modern mathematics, they are not only pedagogical but also provide essential tools in:

- Algebraic number theory (ideal theory and class groups),
- Geometry of numbers (lattice point problems),

- Cryptography (modular arithmetic, RSA, and linear congruential generators),
- Computational mathematics (integer programming and algorithmic complexity).

7.1.6. Scope of This Study

In what follows, we shall:

1. Establish the general solvability criterion.
2. Present constructive methods of obtaining integer solutions.
3. Analyze the parametric form of solutions in two and more variables.
4. Illustrate the applications of linear Diophantine equations in both theoretical and practical contexts.

Thus, the study of linear Diophantine equations forms a cornerstone of number theory: simple to state, rich in structure, and of enduring mathematical significance.

7.2. Diophantine Equations

Equations where we have both integer coefficients and expected solutions are called Diophantine equations.

They are named after Diophantus of Alexandria, a man who studied them in the 3rd century. However, we can also find them in ancient Indian texts, dating back to 800 BC, with important applications in astronomy. The reader has probably encountered Diophantine equations, specifically in geometry, when learning the Pythagorean theorem, x squared plus y squared equals z squared, and appreciated examples in which right triangles had integer side lengths, such as 3, 4, and 5. The ancient Greeks had already addressed this very problem, but it was not yet referred to as Diophantine equations. There were more interesting equations solved in the field of integers, but they were not perceived as a single set of problems, they were solved individually and slowly because they were difficult. It was only in the 20th century that people began to examine the issue comprehensively, and among other things, it eventually became clear that there was no universal approach. We solve

Diophantine equations by type, and it is a challenging field. Therefore, we will focus here on the simplest case, which is linear equations. Since the situation is clear for the equation $ax = b$ (whether the number $\frac{b}{a}$ is an integer or not, mathematicians would say that this is a "trivial" case), it is traditional to start with the second simplest linear equation. See e.g., Mordell [5], Schmidt [7] and Weinberger [9].

Definition 7.1.1. By the term linear Diophantine equation in two unknowns, we mean any equation of the type $ax + by = c$ with unknowns x, y, where $a, b, c \in Z$ and we also require a solution $x, y \in Z$.

Such equations can answer the following questions:

- Can c crowns be paid out using coins of values a and b?
- Can c liters be measured using containers of volumes a and b?
- We let the kids out on their bikes for a lap in the park. One hug takes a minute, the other takes b minutes. When will they meet again?

We have already encountered and solved one equation of this type, namely, when finding the inverse of k modulo n, we had the equation $ax + kn = 1$. It turns out that the method we found there can be easily adapted to the case where, instead of 1 we have c.

Theorem. 7.1.1

Let $a, b, c \in Z$. The linear Diophantine equation $ax + by = c$ has at least one solution, if and only if, c is a multiple of $gcd(a, b)$.

Let us recall that equivalence contains two implications, to which useful terminology is attached. The implication that $gcd(a, b)|\, c$ follows from the existence of a solution can be expressed in the words that divisibility is a necessary condition for the existence of a solution. Without it, the equation cannot be solved, but on the other hand, its truth does not guarantee anything. We therefore also want to know that divisibility is a sufficient condition, which is precisely what the implication in the opposite direction says, that divisibility follows from the existence of a solution. In the proof of the theorem, we must confirm both of these directions. Let us recall the simplest method of knowing that some object solves a given equation: we simply substitute it into the equation.

Proof (instructive): 1) ⇒: Suppose that there exists $x_0, y_0 \in Z$ such that $ax_0 + by_0 = c$. Since $gcd(a,b)$ divides a and b, it must divide c.

2) ⇐: Suppose that $c = k\,gcd(a,b)$ for some $k \in Z$. By Bezout's identity, there exists $A, B \in Z$ such that $Aa + Bb = gcd(a,b)$. Then $kAa + kBb = k\,gcd(a,b)$ or $a(kA) + b(kB) = c$, so the integers $x_0 = kA, y_0 = kB$ solve $ax + by = c$.

The proof provided us with a guide on how to solve the equations.

Example 1: Can 1250 crowns be paid out using coins with values of 6 and 15?

We ask whether it is possible to take x six crowns and y fifteen crowns and add them together to make 1250. Mathematically, it is the equation $6x + 15y = 1250$, where x, y we want to be integers. We know that $gcd(6,15) = 3$, and the number 1250 is not divisible by three, therefore, according to the theorem, it is not possible.

So this didn't work, let's try again.

Example 2: Can 1251 liters be measured using containers with a volume of 6 and 15 liters?

Supplementary question: In which movie was a similar problem solved?

We are looking for a solution to the equation $15x + 6y = 1251$, we have chosen a larger number as a to be ready for the extended Euclidean algorithm. We immediately see that $gcd(15,6) = 3$, and since $\frac{1251}{3} = 417 \in Z$, this problem is solvable.

We need the coefficients of the Bezout identity.

We have $3 = 1 \cdot 15 + (-2) \cdot 6$ or $15 \cdot 1 + 6 \cdot (-2) = 3$. With the coefficients 15 and 6, it is very similar to the equation we are solving, but the right-hand side does not match. We can easily fix this by multiplying, namely multiplying the equation by the number 417. On the left-hand side, we must be careful where we multiply this number, we definitely want to keep the coefficients 15 and 6. Therefore, we get $15 \cdot 417 + 6 \cdot (-834) = 1251$. Comparing it with the given equation, we see that $x = 417$ and $y = -834$ is a solution.

Interpretation in the language of the assignment: First, we add 417 times the content of the fifteen-liter container to the tank, then we remove 834 times the content of the six-liter container, leaving us with 1251 liters. From a practical point of view, it would probably be better to alternate pouring and taking out so that we do not need a tank with a volume of $417 \cdot 15 = 6225$.

Note: According to the theorem and its proof, we should find what the $gcd(a, b)$ is, and if there is a solution, then find some Bezout identity. It is better to do both at once, in one run of Euclid's algorithm. This approach always works; we could consider it an algorithm.

However, there are sometimes opportunities to save work. For example, in our equation, we can notice that all numbers are divisible by three, so we can shorten the equation (which is an equivalent operation even in the world of integers) and solve $5x + 2y = 417$.

If we apply the procedure above to this equation, all calculations will be more pleasant so that it will be worth it.

However, if we are fortunate, we don't even have to calculate anything. We immediately see that $gcd(5, 2) = 1$ (which divides the new right-hand side by 417), and Bezout's identity can be guessed, for example $1 = 1 \cdot 5 + (-2) \cdot 2$. We multiply it by the number 417 and get $417 = 5 \cdot 417 + 2 \cdot (-834)$, from here, we have the solutions $x = 417$ and $y = -834$.

Additional answer: Die Hard 3 (with a Vengeance). Bruce Willis didn't know Diophantine equations, and he almost paid the price for it.

The example showed two things. One is that it is possible to deviate from the algorithm if we know what we are doing. The second thing is that we should wonder how many solutions there are, because the one we found certainly does not satisfy us; we would certainly prefer another one that does not require us to pour out the water that we have laboriously applied before. We have hope because the procedure is based on the Bezout identity, and we have already seen that there are more Bezout expressions.

In mathematical language, we would like to obtain a usable description of the set of all solutions of a given equation.

A reader familiar with linear algebra will certainly not be surprised by the following passages, as they have already encountered something similar for systems of linear equations. Even the proofs will have a similar idea; the only difference is the notation used in the equation. Linearity is a powerful property, and once we have it, many things will follow.

When investigating, it will be very helpful if we frame the situation in a way that allows us to view each solution as a single object and also enables us to utilize tools from linear algebra. We will therefore work with vectors $(x, y) \in Z^2$, for example, in the case of water, we could write that (417 –

834) is a solution to the given equation. The set of all solutions is then a certain subset of the "vector space" Z^2 and we are interested in its properties.

Why did we put the "vector space" in quotes? Because the definition of a vector space requires that the scalars by which we multiply vectors form a field (typically R or C). However, we cannot afford this in Z^2, because the requirement that for (x, y) $\in Z^2$ and the scalar $c, c(x,y) \in Z^2$ forces us to take only integers as the set of scalars. Formally, Z^2 is not a vector space, but a large part of its properties are fulfilled (for example, all algebraic conditions from the definition of a vector space). For a reader who is already familiar with vector spaces (basis, subspaces, etc.), the parallel between vector spaces and what we will discover here is very interesting.

We will choose the notation according to the circumstances; in theoretical considerations, it is better to work with vectors, and in practical problems, it is often more convenient to use the notation $x = \ldots, y = \ldots$

For linear equations, homogeneous equations are much more convenient. This also applies here, so we will make a definition that will allow us to move on to such equations.

Definition 7.1.2. Given a linear Diophantine equation $ax + by = c$, then we define its associated homogeneous equation as $ax + by = 0$.

We will now demonstrate that it is sufficient to be proficient in solving homogeneous equations.

Theorem 7.1.2.

Let $a, b, c \in Z$. Consider the linear Diophantine equation $ax + by = c$. Let $x_p, y_p \in Z^2$ be a solution to it.

The pair $x_2, y_2 \in Z^2$ is a solution of this equation if and only if there exists$x_h, y_h \in Z^2$ such that $(x_0, y_0) = (x_p, y_p) + (x_h, y_h)$ and (x_h, y_h) solve the associated homogeneous equation.

The statement has the form of equivalence and states two essential pieces of information. At the beginning, we somehow obtain one solution to the given equation. The left-to-right implication then says that we can obtain any other solution of it only by adding the solution of the homogeneous equation to that one (we will call them "homogeneous solutions," and we should obtain them more easily, but more on that later). The right-to-left implication then states that nothing can be spoiled by adding homogeneous solutions; we will always obtain a solution to the given equation. Taken together, this means that by adding homogeneous solutions, we get exactly the set of all solutions to the given equation. As usual, we will prove each implication separately and start with the one from right to left (that is, when we add homogeneous solutions, a solution to the given equation is obtained), because it is easier.

Proof (instructive): Let us have some solution (x_p, y_p) of the given equation, so $ax_p + by_p = c$.

1) $\Leftarrow$: Let us assume that $(x_h, y_h) \in Z^2$ solves the associated homogeneous equation. So let us substitute $x_0 = x_p + x_h$ and $y_0 = y_p + y_h$ into the given equation, starting with the left-hand side, and see what comes out of it:

$$ax_0 + by_0 = a(x_p + x_h) + b(y_p + y_h) = (a\,x_p + bx_h) + (ax_h + by_h) = c + 0 = c.$$

Yes, (x_0, y_0) is indeed a solution of the given equation.

2) $\Rightarrow$: Let us assume that (x_0, y_0) solves the given equation. We need to find the vector (x_h, y_h) satisfying two conditions. The algebraic one does not give a choice; to make it work, we must choose $x_h = x_p - x_0$ and $y_h = y_p - y_0$. We claim that this is the vector we are looking for.

Obviously$(x_h, y_h) \in Z^2$ and $(x_0, y_0) = (x_p, y_p) + (x_h, y_h)$. It remains to show that (x_h, y_h) solves the associated homogeneous equation.

We will try to substitute it into its left side, then we will use the fact that (x_p, y_p) and (x_0, y_0) are solutions of the original equation:

$ax_h + by_h = a(x_p - x_0) + b(y_p - y_0) = (ax_p + by_p) - (ax_0 + by_0) = c - c = 0$. Yes, (x_h, y_h) solves the associated homogeneous equation. qed

So, once we have one specific solution (x_p, y_p) of a given Diophantine equation (we call this a particular solution and we can find it by using the Bezout identity), the set of all (integer) solutions can be obtained as follows:

$$\{(x_p, y_p) + (x_h, y_h) : (x_h, y_h) \in Z^2 \wedge ax_h + by_h = 0\}.$$

It remains to figure out how to solve homogeneous equations completely, that is, how to find all integer solutions of such equations. The following theorem will tell us this, but first let us try to think it over. The equation $ax + by = 0$ can be rewritten as $ax = -by$.

We can then easily guess one solution, just give $x = -b$ and $y = a$. Note also that once we have one solution (x_0, y_0), we can get another by multiplying the numbers x_0, y_0 by any integer k, because from $ax_0 = -by_0$ it follows $akx_0 = -bky_0$. There will therefore be infinitely many solutions, for example, all of the form $x = -bk, y = ak$ for $k \in Z$.

The critical question, however, is whether there are others.

Let's imagine for a moment that a, b in the equation $ax = -by$ are incommensurable. The coefficient b divides the left side of ax, but is not a coefficient of a, so it must divide x. In other words, we are looking for a solution for x only among multiples of b. Similarly, we will be looking for a solution for y only among multiples of a. The solutions from the previous paragraph will therefore be (for a, b, not coefficient) the only possible ones. What if a and b are not coefficients? Then there are also solutions other than $(-kb, ka)$. For example, for the equation $4x + 6y = 0$, the previous procedure gives us a solution of the form $x = -6k, y = 4k$ or $(-6k, 4k)$ for $k \in Z$, but we also see a solution $x = 3, y = -2$, which cannot be obtained from the general formula by choosing $k \in Z$. The situation is therefore generally more complicated, but fortunately w, we can determine the coefficient simply by subtracting the coefficients in the equation.

Theorem 7.1.3.

Consider the equation $ax + by = 0$ for $a, b \in Z$. The set of all its integer solutions is

$$\left\{k\frac{b}{\gcd(a,b)}, -k\frac{a}{\gcd(a,b)}, k \in Z\right\}.$$

Proof (informative):

1) First, we verify that the pairs $x_h = k\frac{b}{\gcd(a,b)}, y_h = -k\frac{a}{\gcd(a,b)}$ are indeed solutions from Z.

The integrity follows from the fact that $\frac{b}{\gcd(a,b)}$ and $\frac{a}{\gcd(a,b)}$ are integers, after substituting x_h, y_h into the equation, we immediately get $ax_h + by_h = ak\frac{b}{\gcd(a,b)} - bk\frac{a}{\gcd(a,b)} = 0$.

So it agrees.

2) It remains to show that the solutions given by this rule are all, i.e., that no other exists. Let (x_0, y_0) be a solution to the equation $ax + by = 0$.

Divide it by the number $gcd(a, b)$ and transfer one term to the other side: $\frac{b}{\gcd(a,b)} y_0 = -\frac{a}{\gcd(a,b)} x_0$.

We see that the integer $\frac{b}{\gcd(a,b)}$ must divide $\frac{a}{\gcd(a,b)} x_0$ but $\frac{b}{\gcd(a,b)}$ and $\frac{a}{\gcd(a,b)} x_0$ are coprime numbers, so the number $\frac{b}{\gcd(a,b)}$ must divide x_0.

Therefore, there exists $k \in Z$ such that $x_0 = k\frac{b}{\gcd(a,b)}$, from the equation $by = -ax$ we can easily obtain the corresponding formula for y_0. qed

The question "find a solution to the equation" can be answered in several ways. For example, it is possible to provide a direct description of the set of all solutions, as we did in the text of the theorem. Another (and perhaps more popular) form is to provide a formula that generates such a set using some auxiliary variable; in the theorem, instead of the set, we could write "$x = k\frac{b}{\gcd(a,b)}$, $y = -k\frac{a}{\gcd(a,b)}$ for $k \in Z$. Such a formula is called a general solution.

The attentive reader will have noticed that in the text of the theorem, the minus sign is for y, while in the considerations before the theorem, it was for x. The explanation is easy; we are interested in the set of all solutions, and the placement of the sign does not matter there. If, for example, we were to use the intuitive approach presented before the theorem to solve the equation $3x + 4y = 0$, we would get the set given by the formula $x = -4k, y = 3k$ for $k \in Z$. By choosing $k = 3$, we would then get the specific solution $x = -12, y = 9$ or $(-12, 9)$. If we were to use the formula $(4k, -3k)$ from the theorem, then we would get the same solution by choosing $k = -3$. Thus, both formulas can provide the same solution. They just use a different value of k. It is easy to prove this in general, for all solutions from the set.

Here we encounter the locality of variables, each formula for the solution seems to have its own k, it is not the same letter. It would probably be better to use a different parameter in the second formula instead of k, for example l, then it would be clearer, but it is not necessary. Here we are talking about the set of all solutions, the formula "$x = -4k, y = 3k$ for $k \in Z$ represents the set of pairs that can be obtained by running k through the entire set Z. This object (set of pairs) from the outside does not have any k, it is just a working variable, whose meaning is therefore local and can be replaced by another letter. This alternative formula also creates an infinite set of pairs using its auxiliary variable, and the two sets agree.

So we have a choice when it comes to minus and in practice we usually choose the option that we find more pleasant, for example, for the equation

$10x - 15y = 0$ we get either a solution in the form $(3k, 2k), k \in Z$ or in the form $(-3k, -2k), k \in Z$, we like the first one better.

A similar argument applies to the theorem on the structure of solutions, which did not require any special particular solution (x_p, y_p). If you build your general solution using a certain (x_p, y_p) and your friend decides to start from a different one, he will probably get different specific formulas for the general solution (x, y), but as sets your answers must be equal. In other words, to get to the same specific solution from different particular ones, we will also have to move by different solutions of the associated homogeneous equation, but it will definitely work.

Example 3 (continued from 2): We have already found one solution $x_p = 417, y_p = -834$.

The associated homogeneous equation $15x + 6y = 0$ or $5x = -2y$ has a general solution $x_h = -2k, y_h = 5k \; for \; k \in Z$.

We chose the position of the sign so that two minuses do not meet in the resulting solution for any variable, because we find it aesthetically unpleasant. Feel free to do it the other way around.

By adding, we get the general solution $x = 417 - 2k, y = 5k - 834$ for $k \in Z$.

We can do a test, as usual, by substituting into the equation:

$$15 \cdot (417 - 2k) + 6 \cdot (5k - 834) = 6255 - 30k + 30k - 5004 = 1251.$$

It worked.

If we are interested in solutions from the domain N0, we need that $5k - 834 \geq 0$ and $417 - 2k \geq 0$ or $k \geq \frac{834}{5}$ and $k \leq \frac{417}{2}$.

Such k exist, namely all $k \in N$ satisfying $167 \leq k \leq 208$. We can simply choose some, let's say $k = 200$ and we see that 1251 liters can be obtained, for example, by pouring 17 fifteen-liter containers into the tank and 166 six-liter ones.

We can also choose some criterion among all positive solutions (which brings us to the mathematical field called optimization). Let's imagine, for example, that the weight difference between 15 and 6 liters is not that big for us, but we do not like running for water. We would appreciate a solution that requires the least resources, which means a solution with the smallest number of containers used. Mathematically, this means that we want to minimize $x +$

$y = (417 - 2k) + (5k - 834) = 3k - 417$, but we are only interested in the values of k between 167 and 208. The solution is obviously to choose the smallest possible k, i.e., $k = 167$. We encounter the least trouble if we use $x = 83$ fifteen-liter bottles and one six-liter bottle.

Note: The formula from the theorem on homogeneous equations can be rewritten as $\{k(b', -a'), k \in Z\}$, where the letters a', b' denote the coefficients of the truncated equation. Thus, we have the vector $(b', -a')$, which generates all solutions.

If we were to work in the plane R^2 and let k run through all real numbers, we would get a parametric description of a one-dimensional figure, a line passing through the origin. This is not surprising, we can find integer solutions to the equation $ax + by = 0$ by first solving the equation $ax + by = 0$ in the real domain, obtaining that line, and then checking whether this line intersects any points with integer coordinates. From here follows a similar behavior of equations in R^2 and in Z^2, although the latter is not a vector space (as we have already mentioned). So let's follow that parallel further.

For a parametric line in R^2, we can also choose another direction vector, for example, the opposite vector or $(-b', a')$. This gives us an alternative description of the solution set, exactly as we discussed a moment ago. Now, however, comes the first difference. While in R^2 we can choose other direction vectors, for example $(2b', -2a')$, and they will give us the same line, in Z^2 this is no longer possible because of the restriction $k \in Z$ we would not be able to get from the $(2b', -2a')$ back to $(b', -a')$. Therefore, we will have to be more careful when choosing the generating vector, and we will revisit this later.

In linear algebra, we would also notice that the set of the $\{t(b', -a'), t \in R\}$) is closed under sum and multiplication by a scalar, so it is a vector subspace of R^2, with the vector $(b', -a')$ being its basis. In Z^2, we cannot speak of a subspace and a basis, but the set of all solutions of a homogeneous equation is closed under sum and also under multiplication by an integer scalar. We cannot officially say that this set is one-dimensional, but the feeling is correct; in the real and integer case, we have one degree of freedom represented by the parameter $t \in R$, or $k \in Z$.

Let us now look at the case of the inhomogeneous equation $ax + by = c$, for which we have (for example, using Bezout's identity) obtained a single solution (x_p, y_p). We obtain the set of all solutions by adding this vector to the set of homogeneous solutions. Geometrically speaking, we shift the line (more precisely, the points with integer coordinates from that line) as a whole by the vector (x_p, y_p). The figure shows the case of the equation where $a > 0$ and $b < 0$, then indeed $(-b', a')$ leads upwards to the right. qed

We will translate theoretical knowledge about the solution set into the form of an algorithm.

Algorithm 1

For finding all integer solutions of the equation $ax + by = c$.

0. For example, using the extended Euclidean algorithm, find $gcd(a, b) = Aa + Bb$.

1. If c is not a multiple of $gcd(a, b)$, then there is no solution to the equation.

2. The case $gcd(a, b)$ divides c:

a) Multiply the obtained equality $aA + bB = gcd(a, b)$ by the number $c' = \frac{c}{gcd(a,b)} \in Z$ so that the coefficients a, b are preserved, and you get $a(Ac') + b(Bc') = c$ hence, one particular solution $x_p = Ac', y_p = Bc'$ or the vector $((Ac', Bc')$.

b) Reduce the associated homogeneous equation $ax + by = 0$ by the number $gcd(a, b)$ to the form $a'x + b'y = 0$, which gives the solution $x_h = b'k, y_h = -a'k$ or the pair $(b'k, -a'k)$ for $k \in Z$, or optionally $x_h = -b'k, y_h = a'k$ k or the pair $(-b'k, a'k)$

c) By adding the particular and general homogeneous solutions, you obtain the set of all integer solutions $\{(x_p + kb', y_p - ka'), k \in Z\}$ or $x = x_p + kb', y = y_p - ka'$ for $k \in Z$, or the version with minus by y_h. See e.g., Knuth [3], Medviď [4] and Shaullit [6].

Example 4. Solve the equation $351x + 208y = 143$.
Apply Euclid's algorithm to the numbers 351 and 208.

We get $gcd(351, 208) = 13$, which divides the right side, the equation is solvable.

We also got the expression $351\,3\ +\ 208\,(-5)\ =\ 13$. To have 143 on the right side, we multiply this identity by eleven; on the left side, we want to keep the numbers 351 and 208:

$351\,33\ +\ 208\,(-55)\ =\ 143$. Compared with the given equation, we see the solution $x_p\ =\ 33, y_p\ =\ -55$.

Homogeneous solution: We multiply the equation $351x +\ 208y\ =\ 143$ by 13 and we have $27x +\ 16y\ =\ 0$.

From here $x_h\ =\ -16k, y_h\ =\ 27k$ for $k\ \in\ Z$.

By adding the particular and general homogeneous solutions, we get the general solution of the given equation $x\ =\ 33\ -\ 16k, y\ =\ 27k\ -\ 55$ for $k\ \in\ Z$.

If someone wanted to do it in a set, then the set of all solutions of the given equation is

$$\{(33\ -\ 16k, 27k\ -\ 55) :\ k\ \in\ Z\}.$$

Test: We substitute into the given equation:

$351 \cdot (33\ - 16k) + 208\ \cdot (27k\ -\ 55) = 11583\ -\ 5616k + 5616k - 11440\ = 143$.

If we wanted a solution where $x, y\ \in\ N_0$, then we are out of luck, there are none.

So we have a functional procedure.

Corollary 1.

Consider the linear Diophantine equation $ax + by\ =\ c$. Suppose that c is a multiple of $gcd(a, b)$.

Let $A, B\ \in\ Z$ satisfy $gcd(a, b)\ =\ Aa\ +\ Bb$. Then the set of all solutions of the given equation is

$$\left\{\left(A\frac{c}{gcd(a,b)} + k\frac{b}{gcd(a,b)}, B\frac{c}{gcd(a,b)} - k\frac{a}{gcd(a,b)}\right), k\ \in\ Z\right\}.$$

Some students see this as an alternative solution procedure, they simply memorize these formulas and then plug them in. It works, but experience from exams suggests that it is not very reliable, because it is easy to get something wrong in the formula. We therefore recommend following one of the algorithms above or below. However, for a computer, using the formula is easy and reliable. From a programmer's point of view, the situation is now clear. We have an efficient algorithm for finding Bezout's identity (which we can speed up somewhat by working with negative residues), then we just plug it into the formula, and that's it.

Row Elimination and Euclid's Algorithm 2

Let's go back to example 4a.d for a moment. We exceptionally calculated the entire final row (with a zero on the left) and the numbers -16 and 27, or the vector $(-16, 27)$, appeared there, which is, by an amazing coincidence, exactly the vector that generated the homogeneous solution for us there. It's unlikely to be a coincidence; we clearly have an interesting phenomenon here that warrants closer examination. Let us look at our procedure as a matrix modification.

The algorithm starts with rows $|a|1|0|$ and $|b|0|1|$, which corresponds to the matrix

$$\begin{pmatrix} a & 1 & 0 \\ b & 0 & 1 \end{pmatrix}$$

and it represents the system

$$\begin{aligned} 1\,u + 0\,v &= a \\ 0\,u + 1\,v &= b, \end{aligned}$$

which has a unique solution $u = a, v = b$. We already know that Euclid's algorithm is the same as reducing this matrix to the form

$$\begin{pmatrix} \mathrm{gcd(a,b)} & \mathrm{A} & \mathrm{B} \\ 0 & A_h & B_h \end{pmatrix}.$$

This matrix represents the system

$$A\,u + B\,v = gcd(a,b)$$
$$A_h\,u + B_h v = 0.$$

Since these are equivalent transformations, this system has the same solution as the original one, i.e.,

$$a\,A_p + b\,B_p = gcd(a,b)$$
$$a\,A_h + b\,B_h = 0.$$

The first line confirms that the algorithm yields the Bezout identity. The second equality is interesting, showing that the numbers A_h, B_h solve the associated homogeneous equation. This gives us a vector that generates homogeneous solutions with the formula $x_h = A_h\,k, y_h = B_h\,k$. At this point, however, we do not know whether it generates all of them; this does not follow from this discussion.

Of course, thanks to Theorem 4a.3, we know what the correct generating vectors look like. It is possible to prove that the numbers A_h, B_h are really the numbers from the theorem, but it will take some work. Let's try a different approach, one tailored to our specific situation.

So, let's imagine that someone gave us the numbers A_h, B_h with the idea that they might generate a set of homogeneous solutions. Can we recognize whether this is true?

Lemma 7.1.1.

Let $a, b \in Z$. The vector $(A_h, B_h) \in Z^2$ generates all solutions of the equation $ax + by = 0$ if and only if it solves it and the numbers A_h, B_h are coprime numbers.

Proof (informative): 1) ⇒: If the vector (A_h, B_h) generates a solution, then the equation must also be solved by the numbers $x = A_h\ 1$, $y = B_h\ 1$ or A_h, B_h. It is, therefore, itself a solution.

According to Theorem 7.1.3, there must exist some $k \in Z$ such that $A_h = k\frac{b}{gcd(a,b)} = kb'$ and $y = -k\frac{a}{gcd(a,b)} = -ka'$. So we have $b' | A_h$. However, by the assumption (A_h, B_h) generates all solutions, so it must also generate $(b', -a')$, so $A_h\,|b'$, then $|A_h| = |b'|$, so $|k| = 1$. We can therefore calculate

$gcd(A_h, B_h) = gcd(kb', -ka') = |k| gcd(b', -a') =$
$1.\left(\frac{b}{gcd(a,b)}, \frac{a}{gcd(a,b)}\right) = 1.$

2) ⇐: If (A_h, B_h) solves a homogeneous linear equation, then by Theorem 7.1.3, there must exist some $k \in Z$ such that $A_h = k\frac{b}{gcd(a,b)} = kb'$ and $y = -k\frac{a}{gcd(a,b)} = -ka'$. . We then get $gcd(A_h, B_h) = |k| gcd(b', -a')$.

By assumption $gcd(A_h, B_h) = 1$, so necessarily $k = 1$ or $k = -1$.
If $k = 1$, we get $(A_h, B_h) = (b', -a')$, which is the direct generating vector from Theorem 7.1.3.

If $k = -1$, then we have $(A_h, B_h) = (-b', a')$), which is the other possible generating vector. qed

In our considerations, by adding the multiple of one row to another, we have arrived from the matrix $\begin{pmatrix} a & 1 & 0 \\ b & 0 & 1 \end{pmatrix}$ to the matrix $\begin{pmatrix} \mathrm{gcd(a,b)} & \mathrm{A} & \mathrm{B} \\ 0 & A_h & B_h \end{pmatrix}$.

If we ignore the left column, we see a transition from one square matrix to another, whereas the operation used preserves the determinant. Therefore,

$$det\begin{pmatrix} A & B \\ A_h & B_h \end{pmatrix} = det\begin{pmatrix} 1 & 0 \\ 0 & 1 \end{pmatrix} = 1.$$

So, we have $AB_h - BA_h = 1$. If the number $d \in N$ is a common divisor of A_h and B_h, then according to this formula, it must also divide one, so $d = 1$. We see that there are no common divisors greater than 1, therefore $gcd(A_h, B_h) = 1$ and according to Lemma 7.1.1, the vector(A_h, B_h) generates all homogeneous solutions.

We have thus confirmed that the vector generating a homogeneous solution can be found directly in the table. Another simplification is that we would directly produce c in the first row of the matrix using $gcd(a, b)$. This means that we would go from the matrix

$$\begin{pmatrix} \mathrm{gcd(a,b)} & \mathrm{A} & \mathrm{B} \\ 0 & A_h & B_h \end{pmatrix} \text{to the matrix} \begin{pmatrix} \mathrm{c} & A_p & B_p \\ 0 & A_h & B_h \end{pmatrix}.$$

This matrix represents the system

$$Ap\,u + Bpv = gcd(a,b)$$
$$A_h\,u + B_h\,v = 0.$$

If we substitute the solution $u = a, v = b$, we get

$$aAp + bB_p = c$$
$$aA_h + bB_h = 0.$$

The first row shows that (A_p, B_p) is a particular solution.

We thus obtain an interesting alternative procedure. Given the equation $ax + by = c$, we construct the corresponding matrix and adjust it so that the first column contains the numbers c and 0. In the right part of the rows, we then find the numbers we are looking for to construct the solution.

Algorithm 3

for easy finding of all integer solutions of the equation $ax + by = c$.

0. Apply the extended Euclidean algorithm to the numbers a, b, and negative remainders can also be used. The last row of the table will be $|0|A_h|B_h|$.

1. Create one more row by multiplying the penultimate row (the one with gcd) by a suitable number so that c appears on the left.

If this is not possible, the equation is unsolvable, and the algorithm is finished.

2. If it is possible, the table will add a row $|c|A_p|B_p|$.

The general solution of the equation is

$x = A_p + A_hk, y = B_p + B_hk$ for $k \in Z$, or $x = A_p - A_hk, y = B_p - B_hk$ for $k \in Z$. qed

a, b	A	B
351	1	0
208	0	1
-65	1	-2
13	3	-5
0	16	-27
143	33	-55

Example 5. Let's go back to example 4a.d and solve $351x + 208y = 143$ using the new algorithm. We used negative remainders, which made the process a little faster.

Using the last two lines, we write the general solution $x = 33 + 16k, y = -55 - 27k$ for $k \in Z$.

Of course, the calculations performed are the same as in the previous solution; only the presentation differs. From the point of view of the computer or the programmer, there is no difference between the algorithms.

Nevertheless, it is worth showing this. First, because it is an interesting option for manual calculation (see the next section). Another reason is that Euclid's algorithm works only for two variables, whereas the matrix view can be straightforwardly extended to the case of multiple variables and systems, as seen in Section 8.3.

This is also interesting for the programmer.

7.2.1. Manual Calculation

The previous section presented an interesting option that can make manual calculations significantly more pleasant. The disadvantage is that it is a mechanical procedure that hides what is actually happening. This does not mean that one should not use it, but it is also beneficial to understand what it is based on and what lies behind it, perhaps so that one can deviate from the procedure or abandon it completely when convenient. The reader has a fairly wide range of options to choose from.

He can use the structured procedure according to the first algorithm, speed it up by considering negative residues, and further accelerate it by using the second algorithm. Additionally, other modifications to the procedure are also possible. We will illustrate this with an example.

Example 6: We solve the equation $119x - 273y = -70$.

First, for comparison, we illustrate what the procedure according to Algorithm 1 would look like when we apply the theorems about the structure of the solution.

a, b	A	B
273	1	0
119	0	1
35	1	-2
14	-3	7
7•	7•	-16•
0		

We need to find Bezout's identity for the numbers 119 and -273, but by default, Euclid's algorithm is implemented for $a > b > 0$. Therefore, we use the numbers 273 and 119 in the table and move on to the remainders after division.

We get the identity $7 = 7 \cdot 273 + (-16) \cdot 119$. We adjust it so that it also agrees with the given equation in sign. First, we just rearrange, $119 \cdot (-16) - 273 \cdot (-7) = 7$, then we multiply it by minus ten: $119 \cdot 160 - 273 \cdot 70 = -70$. It worked, so the solution exists. We see the particular solution $x_p = 160, y_p = 70$.

Now, the homogeneous equation. We reduce it by the common divisor 7 to the form $17x - 39y = 0$ and we see the solution $x_h = 39k,\ y_h = 17k$ for $k \in Z$.

Conclusion: The general solution is $x = 160 + 39k, y = 70 + 17k$ for $k \in Z$.

For reasons of economy, someone might still want to choose a better particular solution, for example, by subtracting the homogeneous solution $(39, 17)$ from $(160, 70)$ four times, for the general solution, the formula $x = 4 + 39l, y = 2 + 17l$ for $l \in Z$.

This time, we used a separate letter for "another k" so that it is formally correct. We know that both formulas for the general solution generate the same set, for the sake of interest, we will try to see for ourselves:

$$x = 4 + 39l = 4 + 39(l + 4 - 4) = 4 + 156 + 39(l - 4) = 160 + 39(l - 4),$$

$$y = 2 + 17l = 2 + 17(l + 4 - 4) = 2 + 68 + 17(l - 4) = 70 + 17(l - 4).$$

We see that we can switch between formulas by shifting the parameter by four.

Is it possible to shorten this standard procedure somehow? One possibility is to allow negative remainders in Euclid's algorithm.

However, this does not help here, it would lead to the same calculation. If we look at the table, we will see that $r_{k+1} \leq \frac{1}{2} r_k$ always holds, so our procedure already used optimal negative remainders, which coincidentally were always positive.

A more interesting possibility is offered when we realize that our ultimate goal is not to obtain $gcd(a, b)$, but to express the right-hand side of 70. We know that the number in the left column of the table can be obtained using the coefficients in the other two columns, not only in the last row, but in all of them. So the third row, or $35 = 273\ 1 + 119\ (-2)$, is offered. Multiplying it by minus two and rearranging it gives $119\ 4 - 273\ 2 = -70$, hence $x_p = 4, y_p = 22$. We get nicer numbers from the alternative formula. It probably does not make sense to bother the computer by testing which row it could get to c from, but this can be interesting when calculating by hand, because the "higher" row we use, the smaller the numbers we get for the particular solution.

There's also the question of the sign. Euclid's algorithm works in general for any two inputs. So, we can apply it directly to the coefficients from the equation, in the right order and including the sign. And since we already have a negative number, it makes sense to allow negative remainders. Let's go for it.

a, b	A	B
119	1	0
-273	0	1
199	1	0
-35	2	1
14	7	3
-7•	16•	7•
0		

In the first step, we calculate the remainder after dividing the smaller number by the larger, which is the smaller number. In other words, we subtract the second row from the first zero times, the practical consequence is a change in the order of the numbers. We got $-7 = -273 \cdot 7 + 119 \cdot 16$, multiplying by ten then we get the particular solution. From the computer's point of view, this is ideal, in the worst case we do one extra cycle to adjust the gray and we don't have to deal with the signs at all.

We probably won't come up with anything better within the framework of a structured procedure. Now let's look at the procedure using algorithm 3. Here, it is again possible to work only with positive numbers and add the sign later, or to take negative numbers straight away. Since we want an efficient procedure (otherwise we probably wouldn't use this algorithm), we will take the signed method. We will definitely want to save one step by first writing the number with the larger absolute value into the table. This has one important aspect: we want to read the solution vectors directly from the rows, but the order of the variables is linked to the order of the coefficients in the table. So it is necessary to watch this. The ones in the first two rows best indicate the continuity of the coefficients and columns. I found it useful not to write A and B in the header, but to write the names of the variables that the individual columns relate to directly. Let's go ahead and use the trick of creating a seventieth number using a row starting with 35.

a, b	y	x
-273	1	0
119	0	1
-35	1	2
14	3	7
-7	7	16
0	17	39
-70	2	4

The equation is solvable because we managed to produce a seventy. From the last two lines (thanks to the marked columns), we can easily read that $(x_p, y_p) = (4,2)$ and $(x_h, y_h) = (39,17)k$. That was easy.

If someone wanted to work only with positive numbers because it is easier for them, it is possible; the appropriate sign can be indicated, for example, in the header of the auxiliary columns. As we have already seen in this example, this requires the same steps as the generally more economical method of negative residues.

a, b	$-y$	x
273	1	0
119	0	1
35	1	- 2
14	- 3	7
7	7	-16
0	-17	39
-70	- 2	4

In the table, we then read the correct vectors $(x_p, y_p) = (4,2)$ and $(x_h, y_h) = (39,17)k$.

a, b	x	$-y$
273	0	1
119	1	0
35	- 2	1
14	7	- 3
7	-16	7
0	39	-17
-70	4	- 2

As a final insight into how the algorithm works, we will show that it is not necessary to start by connecting the unit matrix. It is essential to have one 1 in each row of the auxiliary part. Correct labeling of the columns will then ensure the correct interpretation of the results. It is clear that there are many possibilities for adapting the calculation and shaping it according to personal preference.

Generally speaking, the more one intends to deviate from the standard procedure, the better one should understand what the standard procedure actually entails in order to appreciate the impacts of the changes one makes.

Finally, a note. Several times, we searched for more interesting solutions among the existing ones. If we needed a purely N solution for an application, we can see from the formula that there are infinitely many of them.

Exercises

Exercise 1 (routine): Find all solutions $(x, y) \in Z^2$ and $(x, y) \in N_0^2$ for the following Diophantine equations:

$(i) 6x + 9y = 204$; $(iii)\ 10x - 4y = 26$; $(v)\ 819x + 315y = 126$;

$(ii) 10x - 15y = 131$; $(iv) 105x - 75y = 0$; $(vi)\ 65x + 273y = 157$.

Exercise 2: You are given a hundred crowns and told to use them to buy lollipops and candies for a children's day. A lollipop costs five crowns and a candy costs three crowns. What are your options if you do not want to take anything away or subsidize the purchase from your own money?

Exercise 3: You should pay 74 crowns for the parcel based on its weight. The post office only has stamps in the values 4 and 10 left. Will they be able to mark the price for you?

Note: In the old days, parcels were paid for with glued stamps, similar to letters. When I sent a thirty-kilogram parcel of books home from Canada in the spring of 1995, they needed to remove the postage stamps with the Christmas picture and covered the entire two walls of the parcel with them.

Exercise 4: You have a classic scale with two pans and any number of weights weighing 15 or 55 grams. What is the smallest weight you can weigh?

Exercise 5: You have two rods, one 60 dm long and the other 25 dm long. What is the smallest length of fabric that you can measure with them, if you measure along the edge and make lines?

Solution:

1:

(i): $gcd(6,9) = 3$ guess, solve $2x + 3y = 68$: $gcd(3,2) = 1 = 1 \cdot 3 + (-1) \cdot 2$, therefore $gcd(2,3) = 1 = (-1) \cdot 2 + 1 \cdot 3$, after multiplication $2 \cdot (-68) + 3 \cdot 68 = 68$. Solution $(x,y) = (-68 + 3k, 68 - 2k)$ or $x = 3k - 68$, $y = 68 - 2k$ for $k \in Z$. Solution in N_0: $23 \leq k \leq 34$.

(ii): $gcd(10,-15) = 5$ does not divide 131. It has no solution.

(iii): $gcd(10,-4) = 2$ we guess, we solve

$5x - 2y = 13$: $gcd(5,2) = 1 = 1 \cdot 5 + (-2) \cdot 2$, therefore

$gcd(5,-2) = 1 = 1 \cdot 5 + 2 \cdot (-2)$, after multiplication $5 \cdot 13 - 2 \cdot 26 = 13$. Solution $(x,y) = (13 - (-2)k, 26 + 5k)$ or $x = 2k + 13, y = 5k + 26$ for $k \in Z$. Solution in N_0: $k \geq -5$.

(iv): $gcd(105,75) = 15, 7x - 5y = 0$ homogeneous equation. Solution $(x,y) = (5k, 7k)$ or $x = 5k, y = 7k$ for $k \in Z$.

Solution in N_0: $\mathrm{k} \geq 0$.

(v): $gcd(819,315) = 63, 13x + 5y = 2.gcd(819,315) = 63 = 2\ 819 + (-5)\ 315$, therefore 819 4 + 315 (-10) = 126.

Solution $(x, y) = (4 - 5k, -10 + 13k)$ or $x = 4 - 5k, y = 13k - 10$ for $k \in Z$. Solution in N_0: impossible.

(vi): $gcd(65, 273) = 13$ does not divide 157. It has no solution.

2: Equation $5l + 3b = 100. gcd(5, 3) = 1 = 5 \cdot (-1) + 3 \cdot 2$ so $5 \cdot (-100) + 3 \cdot 200 = 100, lp = -100, bp = 200$.

$5l + 3b = 0$ gives $l_h = 3k,\ b_h = -5k$ general solution $l = 3k - 100, b = 200 - 5k$ for $k \in Z$. We want $l, b \geq 0, k \leq 40$ and $k \geq 34$,

Total 8 possibilities
$(2,30), (5,25), (8,20), (11,15), (14,10), (17,5), (20,0), (0,50)$.

3: Equation
$10x + 4y = 74. gcd(10, 4) = 2 = 10.1 + 4\,(-2) so\ 10.37 + 4\,(-74) = 74, x_p = 37, y_p = -74$.
$10x + 4y = 0$ or $5x + 2y = 0$ gives $x_h = -2k, y_h = 5k$, general solution $x = 37 - 2k, y = 5k - 74$ for $k \in Z$. We want $x, y \geq 0$, for example $k = 15$ gives $x = 7$ of ten-crown stamps and $y = 1$ four-crown stamp as solutions.

4: We measure the weight c if we can write $c = 15x + 55y$, where $x, y \in Z$ and negative values mean that we put such weights on the same plate as the object in question. The equation has a solution if $gcd(15, 55)$ divides c, so the smallest weight is 5 grams.

5: We measure the length c if we can write $c = 60x + 25y$, where $x, y \in Z$ and negative values mean that we put on the opposite side. The equation has a solution if $gcd(60, 25)$ divides c, so the smallest length is 5 dm.

7.3. Linear Congruence

Now we move on to the world of integers modulo n. There, too, we sometimes need to solve equations, where equality is taken as equality modulo, or congruence. As usual, the simplest equations are linear ones, but this time we find something interesting even in the equation of one variable, or $ax \equiv b \ (mod\ n)$. See Chionglo [1], Jones [2] and Verdal [8].

Definition. 7.2.1

By the term linear congruence, we denote equations of the type $ax \equiv b \ (mod\ n)$, where $n \in N, a, b \in Z$, and we seek integer solutions x.

What solutions can we expect? We see one thing right away. If x_0 is a solution, then we have a valid congruence $ax_0 \equiv 1 \ (mod\ n)$, in which x_0 can be replaced by any congruent number. We thus obtain an infinite set of solutions $\{x_0 + kn : k \in Z\}$. Is that all? Examples will show that not necessarily.

For example, for the congruence $2x \equiv 6 \ (mod\ 10)$, at first glance, we see the solution $x = 3$, so we have the set of solutions $\{3 + 10k : k \in Z\}$, but by trial and error, we can conclude that $x = 8$ is also a solution. This gives us a new infinite set of solutions $\{8 + 10k : k \in Z\}$, which is disjoint from the previous one. Are there any other solutions? Can we systematically search for them? These are good questions that we will answer right away.

The key to success here is a simple statement.

Fact 7.2.1.

Let $n \in N$. Consider $a, b \in Z$. The number $x_0 \in$ Z solves the linear congruence $ax \equiv b \ (mod\ n)$ if and only if, for some $y_0 \in$ Z, the vector (x_0, y_0) solves the Diophantine equation $ax + ny = b$.

Proof: We prove both directions in equivalence at once.$x_0 \in Z$ is a solution if and only if $ax_0 \equiv b \ (mod\ n)$. This is equivalent to the existence of $y_0 \in Z$ such that $b = ax_0 + y_0 n$. This means, of course, that (x_0, y_0)solves the Diophantine equation $ax + ny = b$. It is therefore sufficient to find all solutions of the equation $ax + ny = b$, after which we ignore the y component and the x component will give us the desired solutions of the linear congruence. This means that the findings from the previous section can be directly transferred to our situation. Results 7.1.1, 7.1.2, and 7.1.3 can be combined in the following statement. qed

Theorem 7.2.2.

Let $n \in N$, consider $a, b \in Z$.

(i) If b is not a multiple of $gcd(a,n)$, then there is no solution to the equation $ax \equiv b \ (mod \ n)$.

(ii) If $gcd(a,n)$ divides b, then the equation $ax \equiv b \ (mod \ n)$ has some solution $xp \in Z$.

Let $n' = \frac{n}{gcd(a,n)}$. The set of all solutions of the linear congruence $ax \equiv b \ (mod \ n)$ is

$$\{x_p + k.n', k \in Z\}.$$

This formula enables the direct calculation of the solution, which is convenient for computers and for students who prefer to memorize formulas. We would rather recommend converting the congruence to a Diophantine equation and solving it in your favourite way, ignoring the y-component.

Example 7: We solve the congruence $45x \equiv 9 \ (mod \ 231)$.

We convert the equation to the Diophantine equation $45x + 231y = 9$. We solve this using algorithm 4a.9. Since we are not interested in y, we do not need to calculate the corresponding values. In fact, we would not even need to include that column, we just wanted to show the reader how it is related to the algorithm from the previous section. Here we need to be a little careful not to ignore the column we need, we would rather label them with the names of the variables.

The solution exists, we get $x = 108 - 77k$ for $k \in Z$.

Note that this corresponds to the formula from the theorem, indeed $77 = \frac{231}{gcd(45,231)}$.

a, b	y	x
231	1	0
45	0	1
6		-5
3		35
0		-77
9		108

Normally, in the world of modulo 231, we could replace the number 108 by shifting it by this value, which is not worth it. However, the formula for the general solution suggests that in this case, shifts of 77 are possible, so it seems that it is possible to offer a more pleasant particular solution than 108. We will also use the possibility of changing the sign of k (not that it is necessary, but addition is in many ways a more pleasant operation than subtraction). We have thus arrived at the alternative formula $x = 31 + 77k$, for which we will do a test:

$45 \cdot (31 + 77k) = 1395 + 3465k = (231 \cdot 6 + 9) + 15 \cdot 231k \equiv 9 + 0k = 9\ (mod\ 231)$.

Incidentally, the more pleasant particular solution $x = 31$ can be obtained directly using the table. As we know, the goal is to get the number 9 in the last row on the left. This can be obtained not only as three times the row of "three," but also by adding the row of "three" and "six." Let us also note that many students prefer a slightly different procedure. They use a structured algorithm, creating the table in its entirety, as they are used to, and ignoring y only comes into play at the stage when the Bezout equation is passed to the solution. Experience with exams suggests that this procedure is more resilient to errors caused by stress. Therefore, we offer it as a more laborious, but perhaps safer option, an example is in Example 9. end

Let's take a closer look at the solutions we obtained, starting from the formula $x = 31 + 77k$. We have a particular solution 31, and as we have already discussed, the rules for substitution in the modulo world immediately imply that numbers of the form $31 + 231k, k \in Z$ are all solutions to the given equation. So, we have one congruent group (the residue class of the number 31) that provides solutions, but clearly not all of them, because according to the formula, there is also a solution $31 + 77 = 108$, which does not lie in the class in question.

By the same reasoning, we conclude that all numbers from the residual class $\{108 + 231k : k \in Z\}$ are solutions, but we still have some more left, because $108 + 77 = 185$ must also be a solution, but it clearly does not belong to the two already formed groups.

Thus, we obtain the third congruent group of solutions $\{185 + 77k : k \in Z\}$. There will be no fourth. If we try to move from the number 185, we get the number $185 + 77 = 31 + 3 \cdot 77 = 31 + 231$, which is already contained in the first considered group. Further addition of the number 77 (or its subtraction) then moves us between the three groups of solutions.

Why do we do this? Let's imagine that we live in the world modulo 231. How many solutions to the equation do we see? Because in this world we consider mutually congruent numbers to be the same thing, when we look at the set $\{31 + 231k\}$ we don't actually see infinitely many numbers, but one, say 31, all the others are just different representatives of the same. The same applies to other residue classes, so we actually see only three different solutions. Their representatives can be selected by generating any three numbers that are not mutually congruent from the general solution $x = 31 + 77k$, say $31, -46$ and 670, and we will have all the solutions from the point of view of the world modulo 231. However, it is customary to take three consecutive numbers from the formula $31 + 77k$, then they are automatically not mutually congruent.

Will something like this work in general? Theorem 7.2.2 says that the solutions come in the form $x_p + \frac{n}{\gcd(a,n)}k, k \in Z$. The first group of solutions is clear, it is the residue class $\{x_p + nk : k \in Z\}$. If $gcd(a,n) > 1$, then the number $x_p + \frac{n}{\gcd(a,n)}$ cannot be congruent to x_p modulo n, so we get the second group of solutions. We continue in this way until we move a total of n away from x_p. It is evident that this happens when we add $gcd(a,b)$ times to the number $\frac{n}{\gcd(a,n)}$.

Since this observation will be useful later, we summarize it in a formal statement.

Theorem 7.2.3.

Let $n \in N$, consider the congruence $ax \equiv b \ (mod\ n)$ for some $a, b \in Z$. Let x_p be some of its particular solutions.

Define the numbers $x_i + \frac{n}{\gcd(a,n)} i$ i for $i = 0, 1, \dots, gcd(a,b) - 1$. The set of all solutions of the given congruence is the union of the sets $\{x_i + kn : k \in Z\}$ for $i = 0, 1, \dots, gcd(a,b) - 1$, these sets are mutually disjoint.

A few chapters later, we will learn that the situation when we divide a set into mutually disjoint parts is called a decomposition. Here, we have obtained a decomposition of the set of all solutions into residue classes, which will soon be very useful to us.

It also follows from our considerations that the range of indices given in the theorem is not the only possible one. We can also take any $gcd(a, n)$ of consecutive integers, calculate the corresponding x_i (then they will not be congruent to each other) and again, we will get a decomposition of the set of all solutions into groups of numbers congruent to these x_i.

This ends the practical part on solving congruences. It was easy; we carried over the main thing from Diophantine equations. Now, however, let's imagine that we have not studied any Diophantine equations and have started to investigate linear congruences straight away. What would we expect? The word "linear" would lead us to similar structural sentences as in the previous section, just in a different language. Namely, we would expect this statement.

Theorem 7.2.4.

Let $n \in N$. Consider the equation $ax \equiv b \ (mod\ n)$ for some $a, b \in Z$, Let x_p be one of its solutions.

A number $x_0 \in Z$ is a solution of the congruence $ax \equiv b \ (mod\ n)$ if and only if there exists $x_h \in Z$ such that $x_0 = x_p + x_h$ and is a solution of the associated homogeneous equation $ax \equiv 0 \ (mod\ n)$.

It is also true that the set of all solutions of a homogeneous equation is, as it were, one-dimensional, given by a single "vector," in this case, the number x_h. We have left the derivation part as Exercise 3. Here, we have again an analogy with linear algebra and systems of linear equations.

Let us show one more example.

Example 8: We solve the equation $30x \equiv 0 \ (mod\ 33)$.

Rewriting it into a Diophantine equation: $30x + 33y = 0$. It is a homogeneous equation, so we just need to shorten it and write the solution straight away: From the equation $10x + 11y = 0$ we get $x_h. = 11k, y_h = -10k$. But we actually solved a congruence, so we ignore y. Conclusion: The general solution of the given congruence is $x = 11k, k \in Z$.

By the way, there are a total of three – see $gcd(30, 33)$ – disjoint residue classes, given, for example, by the representatives $0, 11, 22$.

That was ridiculously easy. Why are we showing it? Because we have a general algorithm for solving equations that should also cope with this example. How will it work then? We should start with Euclid's algorithm.

a, b	y	x
33	1	0
30	0	1
3	1	-1
0	-10	11

We see a row with the correct generator of the homogeneous solution, but we should still generate a particular solution. As a computer, we would be instructed to multiply the penultimate row by the number $\frac{c}{\gcd(a,n)}$, which will give us the necessary information. In this case, we multiply by zero, and a row 0 0 0 will be added to the bottom of the table.

We then get the general solution $x = 0 + 11k, k \in Z$, so the algorithm really arrived at the correct solution, albeit in a somewhat funny way.

Finally, let's think about one possibility to make our lives easier. In the case of Diophantine equations, in a situation where we noticed some common divisors of all numbers a, b, c, we could shorten the equation even before we started the solving process. What if we notice that we could reduce the congruence $ax \equiv b$? Unfortunately, that is not enough.

Lemma 7.2.1.

Let $n \in N$, consider $a, b \in Z$. Suppose that $d \in N$ divides the numbers a, b, n.

Then the number $x_0 \in Z$ solves the equation $ax \equiv b \ (mod\ n)$ if and only if it solves the equation $\frac{a}{d}x \equiv \frac{b}{d}\left(mod\frac{n}{d}\right)$.

Proof: $ax_0 \equiv b \ (mod\ n)$ if and only if there is some $y_0 \in Z$ such that $ax_0 = b + y_0 n$, which is if and only if there is some $y_0 \in$ Z such that

$\frac{a}{d}x_0 = \frac{b}{d} + y_0\frac{n}{d}$, which is if and only if $\frac{a}{d}x_0 \equiv \frac{b}{d}(mod\ \frac{n}{d})$. qed

Since it is easy to forget to truncate the modulus under stress, it is safer to simply omit this and do any truncation only after converting to a Diophantine equation.

Exercises

Exercise 1 (routine): Solve the following congruences:

$(i)\quad 3x \equiv 7\ (mod\ 10);$
$(ii)\ 12x \equiv 0\ (mod\ 20)$
$;+\ (iii)\ 84x \equiv -56\ (mod\ 308);$
$(iv)\quad 3x \equiv 7\ (mod\ 9);$
$(v)\quad 6x \equiv 10\ (mod\ 8);$
$(vi)\ 11x \equiv 0\ (mod\ 40).$

Exercise 2 (routine, instructive): Let $n \in N$, let $a, b \in Z$. Consider some solution x_p of the congruence $ax \equiv b$.

(i) Prove that if $x_0 \in Z$ is a solution to this congruence, then the number $x_h = x_0 - x_p$ solves the congruence $ax \equiv 0\ (mod\ n)$.

(ii) Prove that if $x_h \in Z$ is a solution to the congruence $ax \equiv 0\ (mod\ n)$, then $x_0 = x_p + x_h$ solves the congruence $ax \equiv b(mod\ n)$.

Exercise 3 (routine, instructive): Let $n \in N$, let $a \in Z$. Prove that if x_h solves the congruence ax ≡ 0, then also kx_h solves the congruence $ax \equiv 0$ for any $k \in Z$.

Exercise 4 (routine, instructive): Let $n \in N$ and $c \in Z$, assume that $gcd(c,n) = 1$. Prove that if $x, y \in Z$ satisfy $cx \equiv cy\ (mod\ n)$, then $x \equiv y\ (mod\ n)$.

In fact, we are proving here that in the world of modulo, one can truncate equations, but only by invertible numbers. This corresponds to experience with equations in the real world, where one can also only truncate with invertible (i.e., non-zero) numbers.

Exercise 5 (good, instructive): Let $n \in N$ and $c \in Z$, assume that $gcd(c,n) > 1$. Prove that there exists $x, y \in Z$ such that $cx \equiv cy\ (mod\ n)$, but $x \equiv y\ (mod\ n)$ does not hold.

Solution:

1: (i): $7 = 3x + 10y$, obviously $gcd(3, 10) = 1 = (-3).3 + 1.10$ (can be guessed), we multiply Bezout by seven, $7 = 3.(-21) + 7.10$, so $x = -21$ is the solution.

The equation $3x + 10y = 0$ has the solution $x_h = 10k$, therefore the solution of the given equation is x = −21 + 10k, k ∈ Z. Anyone who wants to can use

$x = 9 + 10k, k \in Z.$

(ii): $12x + 20y = 0$, it is already homogeneous. Evidently $gcd(12,20) = 4$, we shorten, $3x \equiv 0\ (mod\ 5)$ has a solution $x = 5k, k \in Z.$

(iii): $-56 = 84x + 308y$, by Euclid $gcd(308,84) = 28 = (-1).308 + 4.84$, because $\frac{-56}{28} = -2 \in Z$, the equation has a solution, we multiply Bezout by the minus two, $-56 = 84.(-8) + 2.308$, so $x = -8$ is a solution.

The equation $84x + 308y = 0$ is divided by 28 into $3x + 11y = 0$, has the solution $x_h = 11k$, therefore the solution of the given equation is $x = -8 + 11k$, $k \in Z$. Anyone who wants to can use $x = 3 + 11k, k \in Z$.

(iv): since $gcd(3,9) = 3$ and 7 is not a multiple of 3, the equation has no solution.

(v): $10 = 6x + 8y$, obviously $gcd(6,8) = 2 = (-1).6 + 1.8$ (can be guessed), the equation has a solution, since $\frac{10}{2} = 5 \in Z$, we multiply Bezout by the five, $10 = 6.(-5) + 5.8$, therefore $x = -5$ is the solution.

The equation $6x + 8y = 0$ is divided by 2 into $3x + 4y = 0$, has a solution $x_h = 4k$, therefore the solution of the given equation is $x = -5 + 4k$, $k \in Z$. Anyone who wants to, can use $x = 3 + 4k, k \in Z$.

(vi): Since $gcd(11,40) = 1$, the solution set is $x = 40k, k \in Z.$

2: (i): $a(x_0 - x_p) = ax_0 - ax_p \equiv b - b = 0\ (mod\ n).$
(ii) is similar.

3: $a(kx_h) = k\ (ax_h) \equiv k\ 0 = 0\ (mod\ n).$

4: An element c has an inverse element c^{-1} modulo n, thanks to which we can calculate $x = 1x \equiv (c^{-1}\,c)x \equiv c^{-1}\,(cx) \equiv c^{-1}\,(cy) \equiv 1y = y\ (mod\ n).$

Proof by rewriting the congruence is possible, but it contains a problem: For some $k \in Z, cy = cx + kn$, so $kn = c(y - x)$. Since $x - y \in Z$, the number c divides kn. We have $gcd(c, n) = 1$, gives $c \mid k$. Thus $k = cl$ for $l \in Z$, so $c(y - x) = cln$. We want to truncate the integer c, isn't it zero?

Attention! If $n = 1$, then all numbers are congruent to each other, so $x \equiv y \ (mod\ n)$.

Case $n \geq 2$: Since $gcd(0, n) = n \geq 2$, zero does not satisfy the condition $gcd(c, n)$ and therefore indeed $c \neq 0$. We reduce the equation, $y - x = ln$, where $l \in Z$, therefore $x \equiv y \ (mod\ n)$.

5: Let $d = gcd(c, n)$, then $c = dk$ and $n = dm\ for\ k, m \in Z$. Since $d > 1, m < n$. We find $x, y \in Z$ such that $c(y - x)$ is a multiple of n, but $y - x$ is not. However, c already has "a piece of n" in it, so it is enough to arrange that m appears in the difference $y - k$, but not the whole n.

Let us choose, for example, $x = 0$ and $y = m$. Then $0 < y - x = n < m$, so they are incongruent numbers. But $c(y - x) = dkm = k(dm) = kn$.

7.4. Equations in Z_n Spaces

We choose modulus $n \in N$ and move to the finite world Z_n. Even in this world, we restrict ourselves to the simplest type, a linear equation of one variable.

Consider the equation $a * x = b$, where $a, b \in Z_n$. We look for a solution $x_0 \in Z_n$. Starting from the definition of multiplication, we look for x_0 satisfying $(a.x_0)\ mod\ n = b$. Since we have $(a\ x_0)\ mod\ n = b\ mod\ n$, the numbers ax_0 and b have the same remainder when divided by modulus n. It follows that $ax_0 \equiv b\ (mod\ n)$. We have just deduced the following:

If $x_0 \in Z_n$ solves the equation $a * x = b$ in Z_n then x_0 must solve the congruence $ax \equiv b\ (mod\ n)$.

From a practical point of view, this means that there is no point in looking for a solution to the given equation elsewhere than among the solutions of that congruence. So what does it look like with them?

Let's take some solution x_0 of the congruence $ax \equiv b \ (mod\ n)$. Then $(a \cdot x_0)\ mod\ n = b\ mod\ n$, thanks to $b \in Z_n$ therefore $(a \cdot x_0)\ mod\ n = b$. Now we would really like to write $a * x_0$ on the left-hand side, but this operation is defined only for numbers from Z_n. We do not know about the number x_0 and, likely, it does not even lie there.

Fortunately, in the case of congruence, we are allowed to replace numbers with representatives in calculations. The congruence $ax \equiv b \ (mod\ n)$ is therefore solved by the number $r = x_0\ mod\ n \in Z_n$, with which we can get the equality $(a \cdot r)\ mod\ n = b$ in the same way. Here we are already authorized to rewrite as $a * r = b$ in Z_n and we have the solution. We get the following.

If $x_0 \in Z$ is a solution to the congruence $ax \equiv b \ (mod\ n)$, then the remainder after division $r = x_0$ mod n solves the equation $a\,x = b$ in Z_n.

These two findings can be summarized in the following observation:

We obtain the set of all solutions of the equation ax_0 in Z_n by replacing all numbers in the set of all solutions of the congruence $ax \equiv b \ (mod\ n)$ by their remainders after division by n, or by their congruent representatives from the set Z_n.

Theorem 7.2.3 is the starting point. Let us consider one particular remainder class $\{x_i + kn : k \in Z\}$. Since all of these numbers are congruent modulo n, they will also have the same remainders after division. This means that when we go to the remainders, the whole group turns into a single number, for example, the one we get from x_i.

In the previous section, we noted that in the modulo world, we view this infinite set of solutions as a single number, with many possible representatives. By moving to Z_n, this impression becomes reality, the set is now just one number.

This is, of course, true for all groups, so we see that the set of all solutions to the equation Z_n can be obtained by finding the remainders of those numbers xi. Since they were chosen so that they are not congruent modulo n, we get the $gcd(a, n)$ of different solutions from Z_n. Since we had some freedom in choosing the representatives of x_i, the idea is to choose them so that they are from Z_n. Let's try it out, but first, one more thing.

Convention. In the following text, we will use the usual symbols $\cdot$ and $+$ for operations in the space Z_n.

In practical calculations, this is customary, it is much more pleasant to read and write. The problem arises only in theoretical considerations, where it is necessary to carefully distinguish which operations are being discussed. An experienced mathematician will deduce this from the context.

Example 9: We solve the equation $42x = 18$ in Z_{180}.

We rewrite it as the congruence $42x \equiv 18\ (mod\ 180)$, from which we then proceed to the Diophantine equation $42x + 180y = 18$. We typically solve it using a table.

We see information in it to write the general solution, but we are only interested in the variable by 42 or x, the table gives $x_p = 39, x_h = -30$. We get the general solution $x = 39 - 30k, k \in Z$ of the congruence $42x \equiv 18\ (mod\ 180)$.

Before we proceed with the next procedure, we will find a more suitable representative. The smallest non-negative number obtainable from the formula is 9, we also change the sign of k. We will therefore work with the general solution

$$x = 9 + 30k, k \in Z.$$

a, b	y	x
180	1	0
42	0	1
12	1	-4
6	-3	13
0	7	-30
18	-9	39

This set of solutions breaks down into a total of $gcd(42, 180) = 6$ groups (residual classes), namely the groups

$\{9 + 180k : k \in Z\}, \{39 + 180k : k \in Z\}, \{69 + 180l : k \in Z\}, \{99 + 180k : k \in Z\}, \{129 + 180l : k \in Z\}$ and $\{159 + 180k : k \in Z\}$. By the way, we see that another shift would get us to the number 189, which really belongs to the first group, everything agrees.

When we go to the remainders modulo 180 for each group, we get the numbers $9, 39, 69, 99, 129, 159$. These numbers are solutions of the equation $42x = 18$ in Z_{180}.

Conclusion: The set of all solutions of the equation $42x = 18\ (mod\ 180)$ is $\{9, 39, 69, 99, 129, 159\}$. It can also be written as $\{9 +$

$30k : k = 0, 1, \dots, 5\}$, which will come in handy when $gcd(a, n)$ turns out to be a large number.

Note: We deliberately chose a more detailed solution that followed our previous considerations. An experienced solver would probably shorten some of the stages. end

We summarize the procedure in the formal statement.

Theorem 7.3.1.

Let $n \in Z$, consider the equation $ax = b$ in Zn for some $a, b \in Z_n$.

(i) If $gcd(a, n)$ does not divide b, then there is no solution.

(ii) Suppose that $gcd(a, n)$ divides b. Let $x_p \in Z$ solve the congruence $ax \equiv (mod\ n)$, denote $n' = \frac{n}{gcd(a,n)}$.

Let $x_0 = min\{x_p + kn', k \in Z$ and $x_p + kn' \geq 0\}$. Then the set of all solutions of the equation $ax = b$ in Z_n is $\{x_0 + in', i = 0, 1, \dots, gcd(a, n) - 1\}$.

These are the $gcd(a, n)$ of distinct numbers.

Proof (instructive):

(i): Wc will use an indirect proof, i.e., we will prove a variation of this implication. If the equation $ax = b$ had a solution in Z_n, then (see above), it would also be a solution to the congruence $ax \equiv b\ (mod\ n)$.

By Theorem 7.2.2, then $gcd(a, n)$ must divide b.

(ii):

1) First, we will show that all the given numbers are solutions. Let us denote by k_0 the number that generated $x_0 + k_0 n'$.

We know from the previous section that numbers of the form $x_p + kn'\ for\ k \in Z$ solve the congruence $ax \equiv b\ (mod\ n)$. This is also true for the numbers $x_0 + in' = x_p + (k_0 + i)n'$.

According to our previous observations, therefore, their residues $(x_0 + in')\ mod\ n$ solve the equation $ax = b$ in Z_n. It remains to show that the numbers themselves lie in the set Z_n.

By choice, $x_0 \geq 0$, and since $n' > 0$, $x_0 + in' \geq 0$. We still need the restriction from above. We start by observing that $x_0 < n'$.

Otherwise, $x_0 - n' \geq 0$ would hold, which contradicts the definition of x_0 as the smallest nonnegative number of the given type. We can then estimate $x_0 + in' \leq x_0 + (gcd(a,n) - 1)\, n' < n' + (gcd(a,n)) - 1)\, n' = gcd(a,n)$ $.n'$=n, so the numbers are indeed from Z_n.

2) That these are $gcd(a,b)$ of different numbers follows from the fact that $n' \neq 0$, so for $i \neq j$ we have $x_0 + in' \neq x_0 + jn'$. Here, the result of part 1) also plays a role, as these are numbers directly from Z_n, so we do not have to worry that for different i, j we would end up with the same number after forced transition to the residues.

3) Finally, we will show that every solution of the given equation can be found in this list.

So let us consider the solution $\tilde{x} \in Z_n$ of the equation $ax = b$ in Z_n. We have shown that it must solve the congruence $ax \equiv b\ (mod\ n)$.

From Theorem 7.2.2, it follows that $x = x_p + lx_h$ for some $l \in Z$. Then of course

$$\tilde{x} = x_p + ln' = x_0 - k_0 n' + ln' = x_0 + (l - k_0)n' = x_0 + mn',$$

where $m = l - k_0 \in Z$.

The number x is therefore in the correct form; it remains to prove that m lies in the range $0, 1, \ldots, gcd(a,n) - 1$.

We have already deduced that from the choice of x0, we get $x_0 < n'$.

If m were negative, then we would have $\tilde{x} \leq x_0 - n' < 0$, which contradicts $x \in Z_n$. Therefore $m \in N_0$.

We also know that $0 \leq x_0$ and $\tilde{x} \in Z_n$, so $\tilde{x} < n = gcd(a,n) \cdot n'$.

Therefore, we can estimate $mn' = \tilde{x} - x_0 < gcd(a,n)n'$.

Dividing by a positive number n' yields $m < gcd(a,n)$. In summary, the solution $\tilde{x}$ can be written as $x_0 + mn'$, where $m \in \{0, 1, \ldots, gcd(a,n)\}$, so this x is in our list. qed

As usual, it is not necessary to follow the algorithm described in this theorem exactly. For example, it is possible not to introduce a special solution x_0, but to work with the corresponding starting index $k_0 = min\{k: x_p + kn' \geq 0\}$ and to obtain a solution set of the form $\{x_p + kn': k = k_0, k_0 + 1, ..., k_0 + gcd(a,h) - 1\}$.

If you don't like subtracting one, you can try it another way. Consider that the following procedure will give the same result: First, we determine the constant $K = max\{k \in Z: x_p + Kn' < 0\}$ and then we construct the set $\{x_p + kn': k = K+1, K+2, ..., K + gcd(a,h)\}$.

The approach from the theorem would correspond to finding $x_0 = max\{x_p + kn': k \in Z, x_p + kn' < 0\}$ and then using the set $\{x_p + kn': k = +1, 2, ..., gcd(a,h)\}$. The essential thing is that we finally have $gcd(a,n)$ of numbers from the range $0, 1, \ldots, n - 1$. In practice, one simply does what is needed; we are just practicing mathematical notation here. Finally, one more example, the solution will be more concise.

Example 10: Solve the equation $14x = 38$ in Z_{40}.

We translate it as $14x + 40y = 38$ for $x, y \in Z$. We use a convenient algorithm for Diophantine equations. We ignore y, we get the solution set $x = 57 + 20k$. Using $n' = 20$, we find a better representative of 17 and we are ready.

The given equation has solutions $x = 17, 37$.

a, b	y	x
40	1	0
14	0	1
-2	1	-3
0	7	-20
38	-19	57

end.

Exercises

Exercise 6 (routine):
Which of the following equations are solvable in Z_{168}?
a) $25x = 13$; b) $30x = 12$; c) $30x = 15$; d) $16x = 24$.

Exercise 7 (routine):
Solve the following equations in the given Z_n:
(i) $12x = 18$ in Z_{42}; (iii) $10x = 0$ in Z_{35}; (v) $84x = 126$ in Z_{210};
(ii) $9x = 7$ in Z_{20}; (iv) $8x = 10$ in Z_{12}; (vi) $8x = 0$ in Z_{12};

Exercise 8 (good):

Consider the equation $(6 - t)x = 24$ in Z_{40}. For what values of t in the range $0, \ldots, 5$ does this equation have

a) exactly four solutions? b) exactly three solutions? c) exactly five solutions? d) no solutions?

Solution:
6: The condition is gcd $(a, n) | b$.

$a)$: $gcd(25, 168) = 1, 1|13$, yes.
b): $gcd(30, 168) = 6, 6|12$, yes.
c): $gcd(30, 168) = 6, not\ 6|15, no.$
d): $gcd(16, 168) = 8, 8|24, yes.$

7: (i): $18 = 12x + 42y, gcd(42, 12) = 6 = 1 \cdot 42 + (-3) \cdot 12$, we multiply by three, $18 = (-9) \cdot 12 + 3 \cdot 42$, so $x_p = -9$.

The equation $12x + 42y = 0$ reduces to $2x + 7y = 0$, has a solution $x_h = 7k$, therefore the solution of the congruent equation $x = -9 + 7k$.

There is $gcd(42, 12) = 6$ solution in Z_{42}, $x = 5 + 7k$ for $k = 0, 1, \ldots, 5$ or $\{5, 12, 19, 26, 33, 40\}$.

(ii): $9x + 20y = 7, gcd(20, 9) = 1 = (-4) \cdot 20 + 9 \cdot 9$, we multiply to $7 = 9 \cdot 63 + (-28) \cdot 20$, hence $x_p = 63$.

The equation $9x + 20y = 0$ has a solution $Z_{20} = 20k$, therefore $x = 63 + 20k$. Is $gcd(20, 9) = 1$ a solution in Z_{20}, $x = 3$.

(iii): $10x + 35y = 0$, guess $gcd(35, 10) = 5$, reduce to $2x + 7y = 0$, so the solution $x_h = 7k$. Is $gcd(35, 10) = 5$ a solution in Z_{35}, $x = 7k$ for $k = 0, 1, 2, 3, 4$ or $\{0, 7, 14, 21, 28\}$.

(iv): $8x + 12y = 10, gcd(12,8) = 4 = 1 \cdot 12 + (-1) \cdot 8$, since 4 does not divide 10, the equation has no solution.

(v): $84x + 210y = 126, gcd(210,84) = 42 = 1 \cdot 210 + (-2) \cdot 84$, multiply by $126 = (-6) \cdot 84 + 3 \cdot 210$, so

$x_p = -6.$

The equation $84x + 210y = 0$ to $3x + 5y = 0$, has a solution $x_h = 5k$, therefore the solution of the congruent equation is $x = -6 + 5k$. There is $gcd(210,84) = 42$ solution in $Z_{210}, x = 4 + 5k$ for $k = 0,1,\ldots,41$, unofficially $\{4,9,14,19,\ldots,204,209\}$.

(vi): $8x + 12y = 0$, to $2x + 3y = 0$, so $xh = 3k$. There are $gcd(12,8) = 4$ solutions in $Z_{12}, x = 3k$ for $k = 0,1,2,3$ or $\{0,3,6,9\}$.

8: The number of solutions is equal to $gcd(a,n)$, but $gcd(a,n) \mid b$ must hold.

a): We need $gcd(6 - t,40) = 4$, then also $4|24$, this holds for $t = 2$.
b): We need $gcd(6 - t,40) = 3$, this is not possible.
c): We need $gcd(6 - t,40) = 5$, this holds for $t = 1$, but $5|24$ does not hold, so no solution.
d): No solution occurs when $gcd(6 - t,40)$ does not divide 24. Since $40 = 8 \cdot 5$ and $6 - t \leq 6$, the possible gcds are $2,4,5$. Of these, only 5 do not divide 24, the others will have solutions. Conclusion: There will be no solution for $t = 1$.

References

[1] Chionglo, J. F. How I solved the linear congruence 25x = 15 (mod 29). doi:10.13140/RCR2.2.1476856324 (2016).

[2] Jones, G. A., & Jones, J. J. *Congruences.* In J. J. Jones & G. A. Jones, *Elementary Number Theory* (Springer Undergraduate Mathematics Series). London: Springer.

[3] Knuth, D. E. *The Art of Computer Programming*, Vol. 2, Chapter 4. Addison-Wesley. (1981).

[4] Medvid', V. Algorithms of solving the Diophantine equation [electronic]. In *ICERI 2023: 16th International Conference of Education, Research and Innovation, 13–15 November 2023, Seville, Spain* (conference proceedings). (2023).

[5] Mordell, L. J. *Diophantine Equations*. Pure and Applied Mathematics, Vol. 30. Academic Press. ISBN 0-12-506250-8. (1969).

[6] Shallit, J. Origin of the analysis of the Euclidean algorithm. *Historia Mathematica, 21*(4), 401–419. doi:10.1006/hmat.1994.1031 (1994).

[7] Schmidt, W. M. *Diophantine Approximations and Diophantine Equations*. Lecture Notes in Mathematics, Vol. 1467. Berlin: Springer-Verlag. ISBN 3-540-54058-X. (1991).

[8] Verdal, E. Solutions of some classes of congruences. *The Teaching of Mathematics, 9*(1), 41–44. (2006).

[9] Weinberger, P. On Euclidean rings of algebraic integers. In *Proceedings of Symposia in Pure Mathematics, Vol. 24* (pp. 321–332). Providence, RI: American Mathematical Society. doi:10.1090/pspum/024/0337902 (1973).

Chapter 8

Systems of Linear Congruences and Linear Diophantine Equations

Abstract

This chapter focuses on the study of Systems of linear congruences and linear Diophantine equations.
The presented theory is grounded in theoretical knowledge through new definitions, theorems with proofs, and subsequently solved examples and exercises.
One of the main theorems is also the Chinese remainder theorem.
The issue of Decomposition of the module is also treated with a professional explanation and practical demonstrations through examples.
Subsequently, the technique of solving by elimination is presented.
The entire chapter presents two algorithms:
Algorithm for solving the system of congruences $x \equiv b_1 \ (mod\ n_1\ ), x \equiv b_2\ (mod\ n_2), \ldots, x \equiv b_m\ (mod\ n_m)$ for the case where all numbers n_i are not divisible by two

and

Algorithm for solving the system of linear homogeneous equations $A\vec{x}^T = \vec{c}^T$, where A is an $m \times n$ matrix with integer elements, $\vec{c} \in Z_m$ and we expect $\vec{x} \in Z_n$ with practical solutions to the problems.
University students of mathematics have relatively quickly mastered this theoretical knowledge and have successfully solved practical problems related to the given topic.
This chapter is intended for university students of mathematics and also for researchers in the field of algebra and number theory.

Keywords: decomposition, linear Diophantine equation, greatest common divisor, least common multiple, congruence, reduction, elimination, positive integer, Euclidean algorithm, Z_n space

8.1. Introduction

The study of integer solutions to algebraic relations lies at the foundation of number theory. Among the most fundamental objects of this theory are linear congruences and linear Diophantine equations, which describe linear conditions imposed on integers with respect to either equality or modular equivalence. Despite their apparent simplicity, these problems play a central role in arithmetic, algebra, and modern applications such as cryptography, coding theory, and computational complexity.

A linear congruence is an expression of the form

$$ax \equiv b(mod\ m),$$

where $a, b, m \in Z$ with $m > 0$, and one seeks integer solutions xxx. The solvability criterion for this congruence depends on the greatest common divisor gcd (a, m). Specifically, a solution exists if and only if gcd $(a, m) \mid b$. In that case, the congruence admits exactly gcd (a, m) distinct solution classes modulo mmm. Extending this principle, one encounters systems of simultaneous linear congruences, whose compatibility and solution structure are governed by the Chinese Remainder Theorem and generalizations thereof. See e.g., Weisstein [10].

On the other hand, a linear Diophantine equation is a relation of the form

$$a_1x_1 + a_2x_2 + \cdots + a_nx_n = b,$$

with fixed integers $a_1, a_2, \dots, a_n, b$, and the task is to determine integer solutions $(x_1, x_2, \dots, x_n)$. The classical result states that the equation is solvable if and only if gcd $(a_1, a_2, \dots, a_n) \mid b$. When solutions exist, they can be described parametrically in terms of one or more free integer parameters, reflecting the underlying lattice structure of integer solutions.

The two topics are deeply interconnected: solving linear congruences often reduces to solving certain linear Diophantine equations, and conversely, modular methods frequently simplify the analysis of Diophantine problems. This duality illustrates the unifying principle of number theory: the interplay between divisibility, modular arithmetic, and integer linear algebra.

From a theoretical perspective, these problems trace back to the works of Euclid, Diophantus, and later to Gauss, whose *Disquisitiones Arithmeticae* (1801) laid the systematic foundations of congruence theory. In the modern era, they serve not only as introductory examples to algebraic number theory but also as essential tools in computer science and cryptographic protocols, such as the RSA encryption scheme, which relies fundamentally on modular arithmetic and congruence systems.

In this chapter, we shall:

1. Establish the theory of linear congruences, including existence and uniqueness of solutions.
2. Discuss systems of congruences and the Chinese Remainder Theorem.
3. Analyze the structure and solution methods of linear Diophantine equations.
4. Highlight the connections between both frameworks and illustrate them with examples and applications.

Thus, linear congruences and linear Diophantine equations form not only the historical backbone of number theory but also a living field of study with far-reaching implications in pure and applied mathematics.

8.2. Systems of Linear Congruences

Here we will consider the following type of system. Given the moduli $n_1, n_2, \dots, n_m \in N$ and the right-hand sides $b_1, b_2, \dots, b_m \in Z$.

We are looking for integers x such that

$$x \equiv b_1 \ (mod\ n_1),$$
$$x \equiv b_2 \ (mod\ n_2),$$
$$\vdots$$
$$x \equiv b_m \ (mod\ n_m).$$

Let's start with the classics. See e.g., Chionglo [1], Jones [2] and Verdal [8].

Theorem 8.1.1.

Consider the modules $n_1, n_2, \dots, n_m \in N$ and the numbers $b_1, b_2, \dots,$ $b_m \in Z$.

Let x_p be a solution to the system of congruences

$$x \equiv b_1 \ (mod\ n_1),$$
$$x \equiv b_2 \ (mod\ n_2),$$
$$\vdots$$
$$x \equiv b_m \ (mod\ n_m).$$

The number x0 is also a solution of this system if and only if there exists a number $x_h$such that $x_0 = x_p + x_h$ and x_h is a solution of the associated homogeneous system of congruences

$$x \equiv 0 \ (mod\ n_1),$$
$$x \equiv 0 \ (mod\ n_2),$$
$$\vdots$$
$$x \equiv 0 \ (mod\ n_m).$$

The proof differs from the two similar theorems proved earlier only in details, and we leave it to the reader.

As usual, it is sufficient to find one particular solution and then thoroughly investigate the homogeneous equations.

We will start with these, as they are easy.

So, let's consider the system of equations

$$x \equiv 0 \ (mod\ n_1),$$
$$x \equiv 0 \ (mod\ n_2),$$
$$\vdots$$
$$x \equiv 0 \ (mod\ n_m).$$

Each of these equations requires that x be a multiple of the corresponding module, so we are actually looking for such x that are common multiples of all modules n_i. These are exactly numbers of the form $x = k.lcm(n_1, n_2, \ldots, n_m)$ for $k \in Z$. ($lcm(a, b)$ is the least common multiple of a and b).

Homogeneous systems can therefore be easily solved.

From now on, we will limit ourselves to the special case where n_i. are two incomparable, i.e., for $i \neq j$ holds $gcd(n_i, n_j) = 1$. We will immediately obtain a nice solution for the homogeneous equation.

Fact 8 1.2.

Consider the moduli $n_1, n_2, \ldots, n_m \in N$. Assume that these numbers are incomparable.

Then the number $x_0 \in Z$ satisfies the congruence $x \equiv 0\ (mod n_i)$ for all $i = 1, \ldots, m$ if and only if x_0 is a multiple of the number $n_1.n_2. \ldots .n_m$.

We used the relation $lcm(n_1, n_2, \ldots, n_m) = n_1.n_2. \ldots .n_m$.

It remains to figure out how to somehow find one particular solution, which will be more complicated than in the case of Diophantine equations and linear congruences.

We start with the first equation. If x solves it, then it must certainly have the form $x = b_1 + kn_1$. We can find general solutions for the other equations in a similar way, but the problem is that we need one common solution.

So, taking all possible solutions of the first equation $x = b_1 + kn_1$, we should arrange that there is also a particular solution of the second equation among them, in other words, we should arrange that b_2 appears when looking modulo n_2.

The key idea is the following: Since we will have more equations, we do not want the influences in x to mix. More precisely, we have x as the sum of two parts, and so far, it works in such a way that when viewed modulo n_1 the second one is zeroed, and the first one gives the desired b_1. We would like it to work similarly (only in reverse) modulo n_2.

Idea: We will use $x = b_1 + b_2 kn_1$. From the perspective of modulo n_1 everything is still the same, but we are trying to have the number b_2 in the second member when viewed modulo n_2. For this to work, $kn_1 = 1\ (mod\ n_2)$. However, we could have chosen k arbitrarily so far; now, we will choose the correct one, if it exists. Actually, we want k to be the inverse number of to n_1 modulo n_2. Thanks to the assumption of non-coefficients of n_1, n_2, we have it, let's call it x_2.

Now the expression $b_2x_2n_1$ does exactly what we want, modulo n1 gives $(b_2x_2) \cdot n_1 \equiv 0$ and modulo n_2 gives $b_2 \cdot (x_2n_1) \equiv b_2 \cdot 1 = b_2$. Now we still need to arrange for the first expression to behave similarly, but we will arrange that similarly. We get a better version $x = b_1x_1n_2 + b_2x_2n_1$, where x_1 is the inverse of n_2 modulo n_1.

It works nicely; let's consider the case of three equations. Then we build x from three terms, for each we need it to be zero with respect to two moduli, which is easily done by including these moduli. We write down the candidates and clearly write what we expect from them with respect to the different moduli.

n_i	$b_1x_1n_2n_3$	$b_2x_2n_1n_3$	$b_3x_3n_1n_2$
n_1	b_1	0	0
n_2	0	b_2	0
n_3	0	0	b_3

Those zeros already work, no matter what we choose for x_i, so we are free to choose x_i so that the remaining cells in the tableau. For example, if we want $(b_1x_1n_2n_3)\ mod\ n_1 = b_1$, then we need $(x_1n_2n_3)\ mod\ n_1 = 1$. This means that $x_1$should be the inverse of n_2n_3 with respect to the modulus n_1, similarly x_2 should be the inverse of n_1n_3 with respect to the modulus n_2, and x_3 should be the inverse of n_1n_2 with respect to the modulus n_3.

In order for these elements to be found, they must always be n_i non-commutative with the product of the other moduli, so now we see why we limited ourselves to this special case. We have an idea that appears to be effective.

Theorem 8.1.3. (Chinese remainder theorem)

Let $n_1, n_2, \dots, n_m \in N$, $b_1, b_2, \dots, b_m \in Z$. Consider the system of equations

$x \equiv b_1\ (mod\ n_1),$
$x \equiv b_2\ (mod\ n_2),$

.

.

.

$x \equiv b_m \ (mod\ n_m)$.

If all numbers n_i are two-in-one, then this system has a solution $x_0 \in Z$. The set of all solutions is $\{x_0 + kn : k \in Z\}$, where $n = n_1.n_2. \dots .n_m$.

Proof (instructive):

1) First, we show that a solution exists. For $i = 1, \dots, m$ we define $N_i = \frac{n}{n_i}$, so it is the product of all n_j except n_i. Then $gcd(N_i, n_i) = 1$. Therefore, there exists an inverse number x_i to N_i with respect to multiplication modulo n_i. Let $x_0 = \sum_{i=1}^{m} b_i N_i x_i$. We claim that it is a solution to the given system.

Let us choose i. For $i \neq j$ then $n_i | N_j$, therefore $N_j \equiv 0\ (mod\ n_i)$, hence $(b_j N_j x_j) \equiv 0\ (mod\ n_i)$. Subsequently, modulo n_i we obtain $x_0 \equiv b_i N_i x_i \equiv b_i \cdot 1 = b_i\ (mod\ n_i)$.

2) The shape of the set of all solutions follows from the theorem and the fact above.

Instead of stating the set of all solutions, many authors prefer to conclude the Chinese Remainder Theorem by saying that the solution of the system is unique modulo n. How would we prove this?

Let us take another solution y of the system. We can easily see that then $y - x_0$ solves the associated homogeneous system; therefore, by the fact that we have $y - x_0 = kn$ for some $k \in Z$. Then $y \equiv x_0\ (mod\ n)$. qed

The proof of the theorem is given by an algorithm.

Algorithm 4

for solving the system of congruences $x \equiv b_1\ (mod\ n_1), x \equiv b_2\ (mod\ n_2), \dots, x \equiv b_m\ (mod\ n_m)$ for the case where all numbers ni are not divisible by two.

1. Denote $n = n_1.n_2. \dots .n_m$ and $N_i = \frac{n}{n_i}$ for all i.
2. For each i, find the inverse number to N_i with respect to multiplication modulo n_i.

3. Let $x_p = \sum_{i=1}^{m} b_i N_i x_i$. The set of all solutions of the system is $\{x_p + kn : k \in Z\}$. end

Example 11: The theorem is called Chinese because systems of congruences date back to ancient China, sometime in the 3rd century AD.

Perhaps the most famous is the following problem from the classic book "The Mathematical Manual of Master Sun-Tzu" (he was a mathematician, not to be confused with the author of the classic book on military strategy, "The Art of War").

Let's have a certain unknown number of things. If we arrange them in threes, two remain. If we arrange them in fives, three remain. If we arrange them in sevens, two remain. How many things are there?

Translated into modern language, we are looking for a solution to the system of equations $x \equiv 2\ (mod\ 3), x \equiv 3\ (mod\ 5$ and $x \equiv 2\ (mod\ 7)$.

We will use the appropriate algorithm.

We have $n_1 = 3, n_2 = 5, n_3 = 7$, so $n = 3 \cdot 5 \cdot 7 = 105$. We make the complementary products $N_1 = \frac{n}{n_1} = n2 \cdot n_3 = 35, N_2 = \frac{n}{n_2} = n_1 \cdot n_3 = 21, N_3 = \frac{n}{n_3} = n_1 \cdot n_2 = 15$.

Now for each i we need the inverse number to N_i due to multiplication modulo n_i. So we will solve the Diophantine equations

$35x + 3k = 1, 21x + 5k = 1$ and $15x + 7k = 1$.

35		1	0
3	11	0	1
2	1	1	-11
1•	2	-1•	12•
0			

21		1	0
5	4	0	1
1•	5	1•	1•
0			

15		1	0
7	2	0	1
1•	7	1•	-2•
0			

We get the following:

$gcd(35, 3) = 1 = (-1)\ 35 + 12\ 3, so\ 2.35 \equiv 1(mod\ 3)$ and therefore $x_1 = 2$;

$gcd(21, 5) = 1 = 1.21 + (-4)\ 5, so\ 1.21 \equiv 1\ (mod\ 5)$and therefore $x_2 = 1$;

$gcd(15, 7) = 1 = 1.15 + (-2)\ 7, so\ 1.15 \equiv 1(mod\ 7)$ and therefore $x_3 = 1$.

The last two could be estimated without calculation.

We plug this into the formula and get $x_p = 2.2.35 + 3.1.21 + 2.1.15 = 140 + 63 + 30 = 233$. We know that there are shifts of $3.5.7 = 105$ possible so that we can take the better representative 23 as a particular solution.

The solution is $23 + 105k$ for $k \in N_0$ (since we are dealing with a number of things, we have not included negative k). end

The procedure is beautifully algorithmizable; the subroutine for finding inverse numbers can even be called in parallel. The computational time then does not increase much with the number of equations in the system. As usual, shortcuts are offered when calculating manually. The most useful one is that when we work with a specific equation, we temporarily perceive everything from the perspective of the relevant module. This means that we can modify the given equation and also substitute it in calculations. For example, we can replace the second equation with the congruence $x \equiv -2\ (mod\ 3)$. This is probably not worth it, but when we are looking for x_2 for the corresponding term, we can replace the twenty-one in the equation $21x_2 \equiv 1\ (mod\ 5)$ with a congruent representative and solve $1.x \equiv 1\ (mod\ 5)$, which is already worth it.

Example 12: We solve the system $x \equiv 8\ (mod\ 5), x \equiv -1\ (mod\ 6)$, and $x \equiv 14\ (mod\ 7)$.

This example is to remind you that nowhere in the theorem or algorithm was it required that n_i be prime, only pairwise coprime, and there were no requirements at all for the right-hand sides of b_i.

We start by simplifying the equations, each according to the corresponding modulus. So instead of the given one, we will solve the system

$x \equiv 3\ (mod\ 5)$
$x \equiv -1\ (mod\ 6)$ and
$x \equiv 0\ (mod\ 7)$.

Now we also see another simplification, the third term in the solution is multiplied by zero, so we don't have to create it at all. But for practice reasons, we'll do it anyway. To create the individual terms of the solution, we'll use a systematic notation that suits some (like us). For equations for inverse numbers, we'll move on to more pleasant versions.

$x \equiv 3\ (mod\ 5)$	$x \equiv -1\ (mod\ 6)$	$x \equiv 0\ (mod\ 7)$.
3.6.7.? $42x_1 \equiv 1(mod\ 5)$ $2x_1 \equiv 1(mod\ 5)$ $x_1 = 3$	-1.5.7.? $32x_2 \equiv 1(mod\ 6)$ $-x_2 \equiv 1(mod\ 6)$ $x_2 = -1$	0.5.6.? $30x_3 \equiv 1(mod\ 7)\ 2x_3 \equiv 1(mod\ 7)$ $x_3 = 4$

$$x = 3.42.3 + (-1).35.(-1) + 0 = 378 + 35 = 413\,.$$

We have guessed the inverse of x_i, this is often possible, in case of emergency we make tables for the extended Euclidean algorithm. We also have $n = 5 \cdot 6 \cdot 7 = 210$, a better representative is $413 - 210 = 203$.

We get the solution set $x = 203 + 210k\ for\ k \in Z$.

There are two more places in the procedure where we can save work. It often does not matter in principle whether we choose a positive or a negative number for x_i, for example, for x_3 4 and -3 are offered, while there is no such difference between multiplication by three and four.

We can then influence whether the individual terms from which we compose x_p, add up to a large number, or whether they will cancel each other out.

The phase of forming the terms offers another area for simplification. We added 6 7 to the first one to ensure zeroing with respect to the moduli 6 and 7. But the right-hand side of the first equation already added a three, so we only need to add 2 and 7, that is, work with the term $3\,.2\,.7x_1$. We then have the requirement $14x_1 \equiv 1\ (mod\ 5)$. From this point of view, it may be worthwhile to rewrite the second equation in the form $x \equiv 5\ (mod\ 6)$, because then the second term does not have to be $5.5.7x_2$, but $5.7x_2$, where $7x_2 \equiv 1\ (mod\ 6)$, is enough.

The algorithm is therefore quite flexible (when executed manually), especially if we know what it is about. end

The Chinese remainder theorem has many practical applications. For example, it can help speed up calculations in Z_n when n is large and complex. It certainly belongs in the basic arsenal of computer science.

This covers the main points about systems of congruence. As a bonus, we will look at other options for solving systems of congruence. We admit up front that this is nothing particularly practical, but it is quite interesting and also a good warm-up for the brain.

8.3. More about Systems of Congruences

When one hears "systems of linear congruences," one would rather expect equations of the type $a_i x \equiv b_i \ (mod\ n_i)$. However, introducing multiples a_i will dramatically increase the difficulty if we want to solve the equations systematically and at once. That is why this case is not considered. Fortunately, we do not miss it in applications.

A less complicated situation is the situation when we have equations of the type $x \equiv b_i \ (mod\ n_i)$, but the moduli n_i are not two-in-one. The fact that this is a significantly more pleasant case is evidenced, for example, by the fact that there is a reasonable condition for the existence of a solution.

Theorem 8.2.1

$n_1, n_2, \dots, n_m \in N$, $b_1, b_2, \dots, b_m \in Z$. The system of equations. The theorem is called Chinese because systems of congruences date back to ancient China, sometime in the 3rd century AD.

$$\begin{aligned} x &\equiv b_1 \ (mod\ n_1), \\ x &\equiv b_2 \ (mod\ n_2), \\ &\vdots \\ x &\equiv b_m \ (mod\ n_m). \end{aligned}$$

has a solution if and only if for all $i, j \in \{1, \dots, m\}, i \neq j, gcd(n_i, n_j)$ divides $b_i - b_j$.

We have already thought about the fact that the solutions of such a system are unique modulo $lcm(n_1, n_2, \dots, n_m)$. Unlike non-commutative modules, however, we do not have a nice formula for finding the solution. This case is not usually discussed either.

Finally, we will mention two methods that can be used. We will illustrate this with an example for both of them.

Elimination.

Let's go back to example 4d.a, or the system $x \equiv 2\ (mod\ 3), x \equiv 3\ (mod\ 5)\ and\ x \equiv 2\ (mod\ 7)$. Its solution x must satisfy the first equation. What do its solutions look like? We have an algorithm for that.

We immediately see $x_p = 2$, we still need to solve $x + 3y = 0$, which gives $x_h = 3k$. So we have $x = 2 + 3k, k \in Z$.

We will then look for solutions to the other two equations between these numbers, which we will mathematically capture in a familiar way, and we will substitute this x into them.

$$2 + 3k \equiv 3\ (mod\ 5) \Longleftrightarrow 3k \equiv 1\ (mod\ 5)$$
$$2 + 3k \equiv 2\ (mod\ 7) \Longleftrightarrow 3k \equiv 0\ (mod\ 7)$$

We get a system that no longer has any ones at x, but that doesn't matter. We don't want to use the Chinese algorithm anyway. We solve the first congruence in the usual way, and we get $k_0 = 2$ and then $k = 2 + 5l$ for $l \in Z$. We substitute into the third equation:

$$3(2 + 5l) \equiv 0\ (mod\ 7) \Longleftrightarrow 15l \equiv -6\ (mod\ 7) \Longleftrightarrow l \equiv 1\ (mod\ 7).$$

We get $l = 1 + 7m\ for\ m \in Z$. Now we do the back substitution:

$$k = 2 + 5l = 2 + 5(1 + 7m) = 7 + 35m \Longrightarrow x = 2 + 3k = 2 + 3(7 + 35m) = 23 + 105m, m \in Z.$$

We arrive at the same result.

It is clear how we would solve systems with multiple equations; we would simply reduce them one by one. After exhausting the Latin alphabet, we would start to denote the parameters with Greek, Hebrew, etc., letters. Since we solve the congruence at each step, we can handle general systems in this way, including the type $a_i x \equiv b_i\ (mod\ n_i)$ (if they have a solution).

However, this method is not promising for larger systems.

Decomposition of the module.

This method can deal with the case when the modules of the type $x \equiv b_i\ (mod\ n_i)$ are not by pair coprime numbers. Let's use the example of $x \equiv 2\ (mod\ 24), x \equiv 6\ (mod\ 20)$ and $x \equiv 8\ (mod\ 30)$.

Step 1: We decompose all modules into powers of prime numbers, and then we rewrite each congruence

$$x \equiv 2\ (mod\ 3 \cdot 2^3) \Longleftrightarrow \begin{bmatrix} x \equiv 2\ (mod\ 3) \\ x \equiv 2\ (mod\ 2^3) \end{bmatrix}$$

$$x \equiv 6\ (\text{mod } 2^2 \cdot 5) \Longleftrightarrow \begin{bmatrix} x \equiv 6\ (mod\ 2^2) \\ x \equiv 6 (mod\ 5) \end{bmatrix}$$

$$x \equiv 8\ (\text{mod } 2 \cdot 3 \cdot 5) \Longleftrightarrow \begin{bmatrix} x \equiv 8\ (mod\ 2) \\ x \equiv 8 (mod\ 3) \\ x \equiv 8\ (\text{mod } 5) \end{bmatrix}$$

The original system of three congruences is therefore equivalent to a system of seven congruences.

$$\begin{aligned} x &\equiv 2\ (mod\ 2^3) \\ x &\equiv 6\ (mod 2^2) \\ x &\equiv 8\ (mod\ 2) \\ x &\equiv 2\ (mod\ 3) \\ x &\equiv 8\ (mod\ 3) \\ x &\equiv 6\ (mod\ 5) \\ x &\equiv 8\ (mod\ 5) \end{aligned}$$

Step 2: Now we need to consolidate the equations with the same prime number in the base of the power, starting from the highest power. We have three congruences with a modulus based on two. In order for $x \equiv 2\ (mod 2^3)$ to hold, $x = 2 + 8k$ for $k \in Z$.

But then $x = 2 + 4 \cdot (2k) \equiv 2 + 0 \equiv 6\ (mod\ 2^2)$, so these x also satisfy the second equation.

Similarly $x = 2 + 2 \cdot (4k) \equiv 2 + 0 \equiv 8\ (mod\ 2)$ and satisfy the third equation. We see that the equation can replace the first three equations $x \equiv 2\ (mod 2^3)$.

Now we look at the equations with modulus 3. There are two, but $8 \equiv 2\ (mod\ 3)$, so it is actually the same equation. We will consider one of them next.

So far, we have managed to consolidate the first two groups into a system of two equations $x \equiv 2 (mod\ 2^3)\ and\ x \equiv 2\ (mod\ 3)$, where the moduli of two are already coprime.

The last ones are the equations with a modulus of five. The desired x should satisfy $x \equiv 6 \ (mod\ 5)$ and $x \equiv 8(mod\ 5)$, which is not possible at the same time, because $6 \equiv 8 \ (mod\ 5)$ does not hold. Therefore, the system has no solution.

So, let's make a change, in the assignment for the third equation, let's consider $x \equiv 26 \ (mod\ 30)$. How would our solution change? For three equations with a binary base, we would have $x \equiv 26 \ (mod\ 2)$, which can be replaced by the congruence $x \equiv 0 \ (mod\ 6)$, which is satisfied in the case of $x = 2 + 8k$. So we would still replace the binary equations with the equation $x \equiv 2 \ (mod\ 2^3)$.

Since $26 \equiv 2 \ (mod\ 3)$, the second part of step 2 works the same.

And since $26 \equiv 6 \ (mod\ 5)$, in this case, both equations with modulus 5 say the same thing and we just need to take one of them.

Conclusion: The given system, which we converted to 7 equations, can be equivalently replaced by the system $x \equiv 2 \ (mod\ 8), x \equiv 2 \ (mod\ 3), x \equiv 6 \ (mod\ 5)$. Here, the modules are already two-in-one, so we apply the standard algorithm and find a solution.

This procedure is also laborious, and we are glad that in applications, we usually choose the modules ourselves, so that we can ensure mutual coprimality.

Exercises

Exercise 9 (routine, exam): solve the following systems of congruences:

(i) $x \equiv 0 \ (mod\ 3)$ (ii) $x \equiv 4 \ (mod\ 2)$
$x \equiv 1 \ (mod\ 4)$ $x \equiv -4 \ (mod\ 3)$
$x \equiv 2 \ (mod\ 5)$; $x \equiv 4 \ (mod\ 5)$;

(iii) $x \equiv 1 \ (mod\ 7)$ iv) $x \equiv 3 \ (mod\ 5)$
$x \equiv 0 \ (mod\ 9)$ $x \equiv 4 \ (mod\ 4)$
$x \equiv -1 \ (mod\ 11)$; $x \equiv 5 \ (mod\ 3)$

Solution:

9: (i): $n = 60, N_1 = 20$, the inverse in Z_3 is $x_1 = -1$; $N_2 = 15$, the inverse in Z_4 is $x_2 = -1$; $N_3 = 12$, the inverse in Z_5 is $x_3 = -2. x = 0 - 20 - (-1) + 1 - 15 - (-1) + 2 - 12 - (-2) = -63 \equiv$

$57\ (mod\ 60)$. The solutions are $x = 60k - 63$ or, perhaps $57 + 60k$ for $k \in Z$.

(ii): $n = 30, N_1 = 15$, the inverse in Z_2 is $x_1 = 1$; $N_2 = 10$, the inverse in Z_3 is $x_2 = 1$; $N_3 = 6$, the inverse in Z_5 is $x_3 = 1$. $x = 4 - 15 - 1 + (-4) - 10 - 1 + 4 - 6 - 1 = 44 \equiv 14\ (mod\ 30)$. The solutions are $x = 44 + 30k$ or perhaps $14 + 30k$ for $k \in Z$.

(iii): $n = 693, N_1 = 99$, the inverse in Z_7 $x_1 = 1$; $N_2 = 77$, the inverse in Z_9 is $x_2 = 2$; $N_3 = 63$, the inverse in Z_{11} is $x_3 = -4$. $x = 1 - 99 - 1 + 0 - 77 - 2 + (-1) - 63 - (-4) = 351$. The solutions are $x = 351 + 693k$ for $k \in Z$.

(iv): rewriting to $x \equiv 3\ (mod\ 5), x \equiv 0\ (mod\ 4), x \equiv 2\ (mod\ 3)$. $n = 60, N_1 = 12$, the inverse in Z_5 is $x_1 = 3$; N_2 need not be solved; $N_3 = 20$, the inverse in Z_3 is $x_3 = 2$. $x = 3 - 12 - 3 + 0 + 2 - 20 - 2 = 188$. The solutions are $x = 188 + 60k$ or for example $x = 8 + 60k$ for $k \in Z$.

8.4. Systems of Linear Diophantine Equations

This topic is a natural extension of the topic of linear equations with two unknowns. Mathematicians know how to solve systems of linear Diophantine equations, but they use concepts from more advanced matrix theory. Perhaps this is why this topic does not appear in discrete mathematics textbooks. Students must, therefore, turn to books on linear programming or optimization. There, they will find that it is an important and useful topic, and considerable effort is devoted to designing algorithms that can efficiently solve large systems.

As a special treat for readers of this book, we demonstrate here a solution procedure that is elementary (in the sense that it does not use concepts from matrix theory) and generalizes the procedure we used in Section 7.1 in the Chapter 7.

We begin with a single equation with multiple unknowns, $a_1x_1 + a_2x_2 + \cdots + a_nx_n = c$. The procedure for solving the equations $ax + by = c$ has been interpreted in this chapter as matrix adjustments and is based on Bezout's identity search, so it appears that this should suffice.

Recall the highlights from section 4a. In order to find the solution of ax + by = c, we modified the integer row adjustments to the matrix $\begin{pmatrix} a & 1 & 0 \\ b & 0 & 1 \end{pmatrix}$ first on the matrix $\begin{pmatrix} \gcd(a,b) & 1 & 0 \\ 0 & A_h & B_h \end{pmatrix}$ and subsequently on the matrix $\begin{pmatrix} c & A_p & B_p \\ 0 & A_h & B_h \end{pmatrix}$ (if this were possible).

When we then looked at the right-hand side of the matrix (i.e., without the left-hand column), the first row gave the partial vector and the second gave the homogeneous solution generator.

This procedure can also be applied to cases involving multiple numbers. Given the equation $a_1x_1 + a_2x_2 + \cdots + a_nx_n = c$, we can construct matrix

$$\begin{pmatrix} a_1 \; 1 \; 0 \; \ldots \; 0 \\ a_2 \; 0 \; 1 \; \ldots \; 0 \\ \vdots \; \vdots \; \vdots \; \vdots \\ a_n \; 0 \; 0 \; \ldots \; 0 \end{pmatrix}$$

We will refer to the right-hand side as the part that now has the unit matrix (i.e., what remains after omitting the first column). We have shown that the matrix can be converted to a row-reduced matrix by integer row adjustments

$$\begin{pmatrix} \gcd(a_1, a_2, \ldots, a_n) \; A_1 \; A_2 \; \ldots \; A_n \\ 0 \; \vdots \; \vdots \; \vdots \\ \vdots \\ 0 \end{pmatrix}.$$

If the number $\gcd(a_1, a_2, \ldots, a_n)$ divided the right side of the solved equation c, then in the next step, we could get a matrix of the form

$$\begin{pmatrix} c \; A_{p1} \; A_{p2} & \cdots & A_{pn} \\ 0 \; B_{h1} \; B_{h2} & \cdots & B_{hn} \\ 0 \; 0 \; C_{h2} & \cdots & C_{hn} \\ \vdots \; \vdots \; \vdots \; \vdots & & \\ 0 & & \end{pmatrix}.$$

What do the numbers on the right side of the matrix tell us?

The original matrix represented the system

$$\begin{aligned} &1.x_1 + 0.x_2 + \cdots + 0.x_n = a_1 \\ &0.x_1 + 1.x_2 + \cdots + 0.x_n = a_2 \\ &\vdots \\ &0.x_1 + 0.x_2 + \cdots + 1.x_n = a_n \end{aligned}$$

with an unambiguous solution $x_i = a_i$ for all i. Since row adjustments do not change the solution, nor does row multiplication by the constant in the last step, the system given by the last matrix must have the same solution, so after the addition, we get

$$\begin{aligned}
&A_{p1}a_1 + A_{p2}a_2 + \cdots + A_{pn}a_n = c \\
&B_{h1}a_1 + B_{h2}a_2 + \cdots + B_{hn}a_n = 0 \\
&0.\, a_1 + C_{h2}a_2 + \cdots + C_{hn}a_n = 0 \\
&\vdots
\end{aligned}$$

The first line says that $x_1 = A_{p1}, \dots, x_n = A_{pn}$ is the solution to the given equation. Here, it is probably better to switch to the language of vectors, i.e., the vector $(A_{p1}, \dots, A_{pn}) \in Z_n$ solves the given equation.

The second line says that the vector $(B_{h1}, \dots, B_{hn}) \in Z_n$ solves the associated homogeneous equation, as do the third and fourth lines.

Let us denote the individual rows of the right-hand side of the matrix as $\vec{r_1}, \vec{r_2}, \dots, \vec{r_n}$. The equations say that the vector $\vec{r_1}$ is a partial solution of the given equation, while the vectors $\vec{r_2}, \dots, \vec{r_n}$ solve the associated homogeneous equation. It is easy to prove that their arbitrary linear combination $\sum u_i . \vec{r_i}$ solves the associated homogeneous equation.

Now it is interesting for us that the matrix is in row-reduced form, which means that the vectors $\vec{r_2}, \dots, \vec{r_n}$ form a linearly independent set. They therefore generate what we would call an $(n-1)$ −dimensional subspace (superplane) in R^n, but we cannot say that in Z_n, we just imagine it.

One thing we are still missing is to show that these vectors generate all homogeneous solutions. This is no longer as easy; however, it remains true and follows from the knowledge of general systems. To summarize, we have modified a certain matrix into a special form by means of row adjustments of the matrix. The rows of the right-hand side, as vectors $\vec{r_i}$, then allow us to write a general solution of the given equation of the form $\vec{r_1} + \sum_{i=2}^{n} u_i . \vec{r_i}$, where $u_i \in Z$ are the parameters.

The procedure also showed that this is possible precisely when $\gcd(a_1, a_2, \dots, a_n)$ divides c. Thus, we have a nice parallel with the case of two variables. See e.g., Knuth [3], Medvid' [4], [Mordell [5], Shallit [6], Schmidt [7] and Wienberger [9].

Example 13: Solve the equation $6x - 12y + 15z = 45$. Transition to the matrix:

$$\begin{pmatrix} 6 & 1 & 0 & 0 \\ -1 & 0 & 1 & 0 \\ 15 & 0 & 0 & 1 \end{pmatrix} \sim \begin{pmatrix} 6 & 1 & 0 & 0 \\ 0 & 2 & 1 & 0 \\ 3 & -2 & 0 & 1 \end{pmatrix} \sim \begin{pmatrix} 0 & 5 & 0 & -2 \\ 0 & 2 & 1 & 0 \\ 3 & -2 & 0 & 1 \end{pmatrix}$$

$$\sim \begin{pmatrix} 3 & -2 & 0 & 1 \\ 0 & 5 & 0 & -2 \\ 0 & 2 & 1 & 0 \end{pmatrix} \sim \begin{pmatrix} 45 & -30 & 0 & 15 \\ 0 & 1 & -2 & -2 \\ 0 & 2 & 1 & 0 \end{pmatrix} \sim$$

$$\begin{pmatrix} 45 & -30 & 0 & 15 \\ 0 & 1 & -2 & -2 \\ 0 & 0 & 5 & 4 \end{pmatrix}$$

Since we were able to produce the number 45 in the first row and column, the system has a solution.

We get the vectors

$$\vec{r_1} = (-30, 0, 15), \vec{r_2} = (1, -2, -2) \text{ and } \vec{r_3} = (0, 5, 4).$$

The general solution is $(-30, 0, 15) + k(1, -2, -2) + l(0, 5, 4)$

or

$$x = k - 30$$
$$y = 5l - 2k$$
$$z = 15 - 2k + 4l \; for \; k, l \in Z.$$

Test:

$$6(k - 30) - 12(5l - 2k) + 15(15 - 2k + 4l)$$
$$= 6k - 180 - 60l + 24k + 225 - 30k + 60l = 45.$$

For the sake of interest, let us try to verify that the vectors $\vec{r_2}, \vec{r_3}$ are indeed linearly independent in the world of integers. We ask whether it is possible to find integers α, β such that $\alpha(1, -2, -2) + \beta(0, 5, 4) = (0, 0, 0)$ and at least one is not zero. However, the first coordinate gives the equation $1 - \alpha + 0 - \beta = 0$, so necessarily $\alpha = 0$. Substituting into the equation from the second coordinate $-2\alpha + 5\beta = 0, \beta = 0$ comes out as well, so the vectors are indeed linearly independent. end

Note that in linear algebra, vectors are usually thought of as column vectors, but that would not be appropriate here; we would still be transposing. So, we will work with row vectors.

Before going further, it will be useful to see that the equation

$$a_1 x_1 + \cdots + a_n x_n = c$$

can be written in matrix form as

$$(a_1 \ldots a_n) . \begin{pmatrix} x_1 \\ x_2 \\ \vdots \\ x_n \end{pmatrix} = (c).$$

Denote the matrix of the system as A, the vector of unknowns as $\vec{x} = (x_1, x_2, \ldots, x_n)$, and the vector of right-hand sides as $\vec{c} = (c)$, we can capture the equation by writing $A\vec{x}^T = \vec{c}^T$. We obtained the column vectors from the row vectors by transposing. For the vector $\vec{c}$ we did not actually have to, but the reader may guess that we are already setting the stage for the case of multiple equations.

We solved the equation by constructing the matrix $(A^T | E_n)$ which we then reduced by linear adjustments. How will this work for systems of multiple equations? Before we get to them, let's add one bonus.

Elimination

When we introduced the greatest common denominator for multiple numbers, we chose a definition that does it all at once.

First we find $b_2 = \gcd(a_1, a_2)$, then $b_3 = \gcd(b_2, a_3)$, $b_4 = \gcd(b_3, a_4)$, and so on until finally

$$gcd(a_1, \ldots, a_n) = gcd(b_n, a_n).$$

It can be shown that the same number is obtained according to the definition.

Similarly, the Diophantine equation $a_1x_1 + \cdots + a_nx_n = c$ can be solved recursively. In fact, an interesting statement holds:

The numbers $A_1, \ldots, A_n$ solve the equation $a_1x_1 + \cdots + a_nx_n = c$ precisely if there exists $U \in Z$ such that the numbers $U, A_3, \ldots, A_n$ solve the equation

$$\gcd(a_1, a_2)u + a_3x_3 + \cdots + a_nx_n = c$$

and A_1, A_2 solve the equation $a_1x_1 + a_2x_2 = \gcd(a_1, a_2)\mathrm{U}$.

How do we know? Suppose that A_i are the solutions. Then, after plugging them into the given equation and adjusting, we get

$$a_1A_1 + a_2A_2 = c - a_3A_3 - \cdots - a_nA_n. \ (*)$$

This can be interpreted to mean that A_1, A_2 solve the equation

$$a_1x_1 + a_2x_2 = c - a_3A_3 - \cdots - a_nA_n.$$

This means that $\gcd(a_1, a_2)$ must divide the right-hand side or $c - a_3A_3 - \cdots - a_nA_n = U\gcd(a_1, a_2)$ for some $U \in Z$. After adjustment, we get $gcd(a_1, a_2)U + a_3A_3 + \cdots + a_nA_n = c$, i.e., the numbers $U, A_3, \ldots, A_n$ indeed, solve the equation

$$gcd(a_1, a_2)u + a_3x_3 + \cdots + a_nx_n = c.$$

If we then substitute $c - a_3A_3 - \cdots - a_nA_n = U\gcd(a_1, a_2)$ into (*), we get $a_1A_1 + a_2A_2 = gcd(a_1, a_2)U$, so A_1, A_2 solves $a_1x_1 + a_2x_2 = \mathrm{U}.\gcd(a_1, a_2)$.

Now the reverse: Let us have the numbers $U, A_1, \ldots, A_n$ such that A_1, A_2 solve $a_1x_1 + a_2x_2 = \gcd(a_1, a_2)\mathrm{U}$ and $U, A_3, \ldots, A_n$ solve the equation

$$gcd(a_1, a_2)U + a_3A_3 + \cdots + a_nA_n = c.$$

After substitution, we get the formulae

$$a_1x_1 + a_2x_2 = \gcd(a_1, a_2)\mathrm{U},$$

$$gcd(a_1, a_2)U + a_3A_3 + \cdots + a_nA_n = c,$$

which gives

$a_1A_1 + \cdots + a_nA_n = c$ and the numbers $A_1, \ldots, A_n$ therefore solve the given equation.

The statement we have just proved allows us to solve the equation with n unknowns instead of the equation with $n - 1$ unknowns and then, using the solution found, an equation with two unknowns (we already know how to do that). If we are not satisfied with the equation of $n - 1$ unknowns, we can apply the same procedure and obtain the equation of $n - 2$ unknowns, sooner or later, we will arrive at an equation we already know (or it has no solution, which is also a good ending).

So, I wouldn't do it, but I've met people who were comfortable with the reduction. Maybe they're really good at equations with two unknowns. Or they may not be familiar with the matrix procedure.

Let's try it on the equation $6x - 12y + 15z = 45$. Since $gcd(6, -12) = 6$, we should solve the equations $6u + 15z = 45$ and $6x - 12y = 6u$ instead. Naturally, we start with the former. It's simple, it doesn't even pay to try the algorithm. We see that $gcd(6, 15) = 3$, estimate $3 = 1 - 15 + (-2) - 6$, and by multiplying and slightly reorganizing, we arrive at $6 - (-30) + 15 - 15 = 45$. The partial solution $u_p = 30, z_p = 15$ comes out. From the homogeneous equation $6u + 15z = 0$, by truncating it to $2u + 5z = 0$, we see the solution $u_h = 5k$, $z_h = -2k$. Before we put them together, we use the homogeneous solution to find a better partitional representative, $u_p = -30 + 5 - 6 = 0, z_p = 15 - 2 - 6 = 3$, so we continue with the solution $u = 5k, z = 3 - 2k$ for $k \in Z$. We arrive at the second equation, $6x - 12y = 6u$ or $6x - 12y = 30k$. We'll invert it by six and try solving the resulting equation $x - 2y = 5k$ this time with a table for a change. This was embarrassingly simple. In fact, the very second line of the assignment gave gcd. From the last two lines we read the solution $x = 5k + 2l, y = 0 + l$ for $l \in Z$. In summary, we have $x = 5k + 2l, y = l, z = 3 - 2k$ for $k, l \in Z$.

a/b	x	y
-2	0	1
1	1	0
0	2	1
$5k$	$5k$	0

That looks suspiciously nice.

Test: $6(5k + 2l) - 12l + 15(3 - 2k) = 30k + 12l - 12l + 45 - 30k = 45$.

So really, that's the solution. But are they all solutions? One way to find out is to take a general solution that we obtain in a way we trust, from the previous example, and attempt to derive it using the new formula. So, to avoid confusing the parameters, let us consider the solutions $x = K - 30, y = 5L - 2K, z = 15 - 2K + 4L$ for some K, L ∈ Z. Can we obtain it using the new formulas? It means solving the system of equations

$$K - 30 = 5k + 2l$$
$$5L - 2K = l$$
$$15 - 2K + 4L = 3 - 2k$$

with parameters $K, L \in Z$ and unknowns $k, l \in Z$. The second and third equations give $k = K - 2L - 6 \text{ and } l = 5L - 2K$, which are integers, substituting them into the first equation confirms that they do indeed solve the system, so everything is fine.

A curious reader might still try the opposite view, i.e., whether all new solutions can be obtained using the previous ones. This means that the roles of parameters and unknowns are reversed when solving the system. It is then confirmed that both formulas define the same subset of Z^3.

11 Systems of equations:

Consider a general system of m equations with n unknowns

$$a_{1,1}x_1 + a_{1,2}x_2 + \cdots + a_{1,n}x_n = c_1$$
$$a_{2,1}x_1 + a_{22}x_2 + \cdots + a_{2,n}x_n = c_2$$
$$\vdots$$
$$a_{m,1}x_1 + a_{m,2}x_2 + \cdots + a_{m,n}x_n = c_m$$

Typically, we have $m \leq n$.

Given the procedure for a single equation, one would expect that it would now want to pile the coefficients of the individual equations into columns (this would be the "left part" of the working matrix), corresponding to the A^T as we thought of it for the case of one equation. To this, we glue a unit matrix of suitable size (the "right part" of the working matrix) and then reduce.

We get a matrix in a row-reduced form. To help the reader's imagination, we have shown one such typical example on the right, which involves two equations with four unknowns. We have also visually separated the left and right parts. What do we know about the left part? The first column will have a number at the top, namely $\pm\, gcd(a_{1,1}, \ldots, a_{1,n})$ and zeros underneath, and in the next columns, it is not clear what the non-zero numbers mean.

$$\left(\begin{array}{cc|cccc} * & * & * & * & * & * \\ 0 & * & * & * & * & * \\ 0 & 0 & * & * & * & * \\ 0 & 0 & 0 & * & * & * \end{array}\right)$$

Since we have n rows and the left part of the matrix (the one representing each equation) has m columns, it will contain (if $m \le n$) at least $n - m$ zero rows. If the analogy with the single equation case holds, then on these rows we should look for vectors that generate homogeneous solutions on the right-hand side. The calculus seems to agree with intuition.

We start with the space Z^n and ideally, each equation takes one "dimension" or one parameter from the solution, so that the general solution can be expected to have $n - m$ parameters (assuming that the equations are not somehow coupled, this would add zero rows).

We will then look for partial solutions in those rows that do not have zeros on the left-hand side. In the first column, we should probably (see the procedure for one equation) produce c_1 instead of gcd, by cleverly multiplying the first row, if we can. But in the second column (representing the second equation), it will not be so easy. We should probably make c_2 by multiplying the rows. Since the first row of the matrix has already served to produce c_1 in the first column, and the third (and subsequent) rows have zero in the second column, it is evident that this c_2 will be produced using the second row, but it is not immediately clear how this will be done.

It can be worked out intuitively, but we prefer to take our advice. As usual, we will work with the matrix notation of the system.

Algorithm 5:

for solving the system of linear homogeneous equations $A\vec{x}^T = \vec{c}^T$, where A is an $m \times n$ matrix with integer elements, $\vec{c} \in Z_m$ and we expect $\vec{x} \in Z_n$.

1. Create a matrix $(A^T|E_n)$ and convert it by integer row adjustments to row-stair form $(D|S)$.

Let $N \in N$ be the largest among the row numbers that are not identically zero (if there are any).

2. Using integer multiplication of the first N rows of the matrix $(D|S)$, modify it into the form (B|R) such that for each $j = 1,..,m$, the sum of the elements in column j of matrix B is equal to c_j.

If this is not possible, the given system has no solution.

3. Assume that you have succeeded in step 2.

Let $\vec{r_i}$ be the rows of the matrix R. Denote by $\overrightarrow{x_p} = \sum_{i=1}^{N} u_i\vec{r_i}$.

The general solution of the system $A\vec{x}^T = \vec{c}^T$ is then given by $\overrightarrow{x_p} + \sum_{i>N} u_i\vec{r_i}$ for $u_i \in Z$. end

Consider that when there is only one equation, $m = 1$, this algorithm agrees with the procedure presented in the previous section. The proof of the algorithm's correctness is a bit longer; we will dispense with it in the bonus section.

The second step remains a bit unclear. How do we know when the prescribed modification is possible and when it is not? The key here is that the matrix D is in a stepwise form. First, we will show an example. Let us agree that the word "leading row member" refers to the first non-zero element of a row of the matrix, counted from the left (if one exists).

Example 14: Solve the system

$$10w + 14x + 16y + 18z = 4$$
$$w - 10x + y + 3z = -5$$
$$2w - x + 3y + 4z = -1$$
$$9w + 13x + 16y + 17z = 0$$

The matrix of the system is $A = \begin{pmatrix} 10 & 14 & 16 & 18 \\ 1 & -10 & 1 & 3 \\ 2 & -1 & 3 & 4 \\ 9 & 13 & 16 & 17 \end{pmatrix}$, according to the algorithm we should work with the matrix

$$\left(\begin{array}{cccc|cccc} 10 & 1 & 2 & 9 & 1 & 0 & 0 & 0 \\ 14 & -10 & -1 & 13 & 0 & 1 & 0 & 0 \\ 16 & 1 & 3 & 16 & 0 & 0 & 1 & 0 \\ 18 & 3 & 4 & 17 & 0 & 0 & 0 & 1 \end{array}\right).$$

Step 1: We need to modify the matrix to row-staircase form using integer row operations.

We start by reducing the first column. We choose the smallest number (in absolute value) as the pivot; in this case, it is 10 in the first row. We subtract the appropriate multiples from the other rows to get as many of the smallest numbers in the first column as possible, allowing negative numbers. We get

$$\left(\begin{array}{cccc|cccc} 10 & 1 & 2 & 9 & 1 & 0 & 0 & 0 \\ 4 & -11 & -3 & 4 & -1 & 1 & 0 & 0 \\ -4 & -1 & -1 & -2 & -2 & 0 & 1 & 0 \\ -2 & 1 & 0 & -1 & -2 & 0 & 0 & 1 \end{array}\right).$$

Now the smallest number is -2 in the fourth row. We can see that it divides all the other elements of the first column, so the first stage ends here. We will use the fourth row to produce the zeros in the first column and then replace it with the first. There is no need to multiply it by minus one. We don't mind negative numbers. The matrix comes out

$$\left(\begin{array}{cccc|cccc} -2 & 1 & 0 & -1 & -2 & 0 & 0 & 0 \\ 0 & -9 & -3 & 2 & -5 & 1 & 0 & 2 \\ 0 & -3 & -1 & 0 & 2 & 0 & 1 & -2 \\ 0 & 6 & 2 & 4 & -9 & 0 & 0 & 5 \end{array}\right).$$

Now we move to the second column. The smallest number there (not considering the first row) is -3 in the third row. We will make zeros in the second and fourth rows and then replace the third with the second. And in order to have the diagonal something positive on the diagonal, we'll multiply it by minus one just for the fun of it. You should get

$$\left(\begin{array}{cccc|cccc} -2 & 1 & 0 & -1 & -2 & 0 & 0 & 1 \\ 0 & 3 & 1 & 0 & -2 & 0 & -1 & 2 \\ 0 & 0 & 0 & 2 & -11 & 1 & -3 & 8 \\ 0 & 0 & 0 & 4 & -5 & 0 & 2 & 1 \end{array}\right)$$

There is no pivot in the third column, so we are left with at most three non-zero rows in the left part of the matrix. Actually, we see that there are exactly three. In other words, there is one generating vector for the homogeneous solution. In order to make this step, we move to the fourth column and produce a zero in the fourth row using the third.

$$\left(\begin{array}{cccc|cccc} -2 & 1 & 0 & -1 & -2 & 0 & 0 & 1 \\ 0 & 3 & 1 & 0 & -2 & 0 & -1 & 2 \\ 0 & 0 & 0 & 2 & -11 & 1 & -3 & 8 \\ 0 & 0 & 0 & 0 & 17 & -2 & 8 & -15 \end{array}\right)$$

We have $N = 3$.

Step 2: We need to produce the correct sums in the columns.
To make the first column add up to 4, multiply the first row by -2:

$$\left(\begin{array}{cccc|cccc} 4 & -2 & 0 & 2 & 4 & 0 & 0 & -2 \\ 0 & 3 & 1 & 0 & -2 & 0 & -1 & 2 \\ 0 & 0 & 0 & 2 & -11 & 1 & -3 & 8 \\ 0 & 0 & 0 & 0 & 17 & -2 & 8 & -15 \end{array}\right).$$

To make the second column add up to -5, we multiply the second row by -1. Because we have the leading member of the row there, it has to be zero in the first column, so this multiplication does not mess up what we just created in the first column.

$$\left(\begin{array}{cccc|cccc} 4 & -2 & 0 & 2 & 4 & 0 & 0 & -2 \\ 0 & -3 & -1 & 0 & 2 & 0 & 1 & -2 \\ 0 & 0 & 0 & 2 & -11 & 1 & -3 & 8 \\ 0 & 0 & 0 & 0 & 17 & -2 & 8 & -15 \end{array}\right).$$

The sum of the third column cannot be influenced. The first two rows are already set, the only chance to affect something is to have a leading member in the third row, but it is not there. We can only check if the third column has the correct sum. Yes, it comes out -1, we are lucky.

In the fourth column, we have the leading member of the third row, so we have a chance to set it. To make the fourth column add to zero, we multiply the third row by -1:

$$\left(\begin{array}{cccc|cccc} 4 & -2 & 0 & 2 & 4 & 0 & 0 & -2 \\ 0 & -3 & -1 & 0 & 2 & 0 & 1 & -2 \\ 0 & 0 & 0 & -2 & 11 & -1 & 3 & -8 \\ 0 & 0 & 0 & 0 & 17 & -2 & 8 & -15 \end{array}\right).$$

Step 3: We have $N = 3$ non-zero rows, so $\overrightarrow{x_h} = (17,-2,8,-15)$ and $\overrightarrow{x_p} = (4,0,0,-2) + (2,0,1,-2) + (11,-1,3,-8) = (17,-1,4,-12)$.

The general solution of the system is $(17,-1,4,-12) + u(17,-2,8,-15)$ or

$$w = 17 + 17u,$$
$$x = -1 - 2u,$$
$$y = 4 + 8u,$$
$$z = -12 - 15u, u \in Z.$$

The test confirms that these numbers do indeed solve all four equations (we really did). end

The procedure we saw in step 2 can be formalized.

0. Set $j = 1$.

1. If there is a leading member of a row in column j, multiply the row in question by an integer so that the sum of column j is equal to c_j . If this does not work, the system has no solution, and the procedure ends.

If no leading member of a row is found in column j, verify that the sum of its elements is c_j. If this is not satisfied, the system has no solution and the procedure terminates. If $j < m$, increase j by one and repeat step 1.

It is not difficult to show that this procedure can correctly detect when the production of the matrix $(B|R)$ is possible and when it is not.

One more example.

Example 15: Consider a system of equations

$$15s + 24t - 12u + 30v = 6$$
$$10s - 4t + 26u + 14v = 12.$$

We have the matrix of the system $A = \begin{pmatrix} 15 & 24 & -12 & 30 \\ 10 & -4 & 26 & 14 \end{pmatrix}$.

We construct the working matrix and reduce it.

$$\left(\begin{array}{cc|cccc} 15 & 10 & 1 & 0 & 0 & 0 \\ 24 & -4 & 0 & 1 & 0 & 0 \\ -12 & 26 & 0 & 0 & 1 & 0 \\ 30 & 14 & 0 & 0 & 0 & 1 \end{array}\right) \sim \left(\begin{array}{cc|cccc} 3 & 36 & 1 & 0 & 1 & 0 \\ 0 & 48 & 0 & 1 & 2 & 0 \\ -12 & 26 & 0 & 0 & 1 & 0 \\ -6 & 92 & 0 & 0 & 3 & 1 \end{array}\right) \sim \left(\begin{array}{cc|cccc} 3 & 36 & 1 & 0 & 1 & 0 \\ 0 & 48 & 0 & 1 & 2 & 0 \\ 0 & 170 & 4 & 0 & 5 & 0 \\ 0 & 164 & 2 & 0 & 5 & 1 \end{array}\right)$$

$$\sim \left(\begin{array}{cc|cccc} 3 & 36 & 1 & 0 & 1 & 0 \\ 0 & 48 & 0 & 1 & 2 & 0 \\ 0 & -22 & 4 & -4 & -3 & 0 \\ 0 & 20 & 2 & -3 & -1 & 1 \end{array}\right) \sim \left(\begin{array}{cc|cccc} 3 & 36 & 1 & 0 & 1 & 0 \\ 0 & 8 & -4 & 7 & 4 & -2 \\ 0 & -2 & 6 & -7 & -4 & 1 \\ 0 & 20 & 2 & -3 & -1 & 1 \end{array}\right)$$

$$\sim \left(\begin{array}{cc|cccc} 3 & 36 & 1 & 0 & 1 & 0 \\ 0 & 0 & 20 & -21 & -12 & 2 \\ 0 & -2 & 6 & -7 & -4 & 1 \\ 0 & 0 & 62 & -73 & -41 & 11 \end{array}\right) \sim \left(\begin{array}{cc|cccc} 3 & 36 & 1 & 0 & 1 & 0 \\ 0 & -2 & 6 & -7 & -4 & 1 \\ 0 & 0 & 20 & -21 & -12 & 2 \\ 0 & 0 & 62 & -73 & -41 & 11 \end{array}\right)$$

$$\sim \left(\begin{array}{cc|cccc} 3 & 36 & 1 & 0 & 1 & 0 \\ 0 & -2 & 6 & -7 & -4 & 1 \\ 0 & 0 & 20 & -21 & -12 & 2 \\ 0 & 0 & 2 & -10 & -5 & 5 \end{array}\right) \sim \left(\begin{array}{cc|cccc} 3 & 36 & 1 & 0 & 1 & 0 \\ 0 & -2 & 6 & -7 & -4 & 1 \\ 0 & 0 & 0 & 79 & 38 & -48 \\ 0 & 0 & 2 & -10 & -5 & 5 \end{array}\right)$$

Then we will switch the last two lines.

Now comes the editing phase. First, in the first column, we need the result 6, so we multiply the first row by 2. We move on to the second column and multiply the second row so that the second column gives the sum of 12. This is also possible.

$$\left(\begin{array}{cc|cccc} 3 & 36 & 1 & 0 & 1 & 0 \\ 0 & -2 & 6 & -7 & -4 & 1 \\ 0 & 0 & 2 & -10 & -5 & 5 \\ 0 & 0 & 0 & 79 & 38 & -48 \end{array}\right)$$

$$\left(\begin{array}{cc|cccc} 6 & 72 & 2 & 0 & 2 & 0 \\ 0 & -2 & 6 & -7 & -4 & 1 \\ 0 & 0 & 2 & -10 & -5 & 5 \\ 0 & 0 & 0 & 79 & 38 & -48 \end{array}\right)$$

$$\left(\begin{array}{cc|cccc} 6 & 72 & 2 & 0 & 2 & 0 \\ 0 & -60 & 180 & -210 & -120 & 30 \\ 0 & 0 & 2 & -10 & -5 & 5 \\ 0 & 0 & 0 & 79 & 38 & -48 \end{array}\right).$$

By the way, we see that the condition for solvability is that c_1 is a multiple of three and $c_2 - 72$ is even, so c_2 must be even.

We come to the last stage of the solution. The first two lines are non-zero on the left-hand side, so the partial solution is

$$\overrightarrow{x_p} = (2,0,2,0) + (180,-210,-120,30) = (182,-210,-118,30)$$

and the general solution of the associated homogeneous equation is

$$\overrightarrow{x_h} = k(2,-10,-5,5) + l(0,79,38,-48).$$

We have the solution to our system

$s = 182 + 2k$
$t = -210 - 10k + 79l$
$u = -118 - 5k + 38l$
$v = 30 + 5k - 48l$
for $k, l \in Z$. end

Note: Especially in the case of large matrices, it is useful to know that it was not necessary to create a stepped shape in the right-hand part of the matrix. We have recognized that the vectors generating homogeneous solutions will be independent, but this must always be true. This is because originally, we had a unit matrix of rank n in the right part of the matrix, which will not change even after row operations. Therefore, the rows of the right-hand part of the matrix must be linearly independent, whether we look at them in any stage of the adjustment.

It is therefore possible to produce a stepped shape only in the columns of the left-hand part of the matrix, which can be a significant time-saver.

In the previous example, we carried the whole process through to the end, because it is quite nice to have simpler vectors as generators, but sometimes we don't feel like it.

Exercises

Exercise 10 (routine): find general solutions for the following equations:

(i) $21x + 28x - 14y = 35$; (iii) $3a + 5b + 7c = 9$,
(ii) $8x - 12y + 28z = 18$; (iv) $2r - 3s + 4t - 5u + 6v = 7$.

Exercise 11 (routine): find general solutions for the following systems of equations:

(i)
$25x - 15y = 10$
$4x + 6y = 52;$

(ii)
$9x + 6y = 5$
$6x - 4y = 10;$

(iii)
$3r + 5s - 7t + 9u = 2$
$3r - 5s + 7t - 9u = 3$
$2r + 4s + 6t + 8u = 4;$

(iv)
$6u - 9v + 9w = 12$
$2u - 4v + 5w = 13;$

(v)
$3x + 4y + 5z = 6$
$4x + 5y + 6z = 7$

Solution:

Note: Because there are multiple possible approaches to matrix reduction, your matrix may be different. The bottom line is that there are several different ways to reduce your matrix.

Your vectors should generate the same solution space.

10: (i): $\left(\begin{array}{c|ccc} 21 & 1 & 0 & 0 \\ 28 & 0 & 1 & 0 \\ -14 & 0 & 0 & 1 \end{array}\right) \sim \left(\begin{array}{c|ccc} 7 & 1 & 0 & 0 \\ 0 & 0 & 1 & 2 \\ 0 & 2 & 0 & 3 \end{array}\right) \to \left(\begin{array}{c|ccc} 35 & 5 & 0 & 5 \\ 0 & 0 & 1 & 2 \\ 0 & 2 & 0 & 3 \end{array}\right)$

$(5,0,5) + k(0,1,2) + l(2,0,3)$ or $x = 5 + 2l, y = k, z = 5 + 2k + 3l$ for $k, l \in Z$.

By the way, $gcd(21, 28, -14) = 7$.

(ii): $\left(\begin{array}{c|ccc} 8 & 1 & 0 & 0 \\ -12 & 0 & 1 & 0 \\ 28 & 0 & 0 & 1 \end{array}\right) \sim \left(\begin{array}{c|ccc} 4 & 2 & 1 & 0 \\ 0 & -3 & -2 & 0 \\ 0 & -5 & -1 & 1 \end{array}\right)$

Cannot $4 \to 18$, no solution exists.

(iii): $\left(\begin{array}{c|ccc} 3 & 1 & 0 & 0 \\ 5 & 0 & 1 & 0 \\ 7 & 0 & 0 & 1 \end{array}\right) \sim \left(\begin{array}{c|ccc} 1 & 2 & -1 & 0 \\ 0 & -5 & 3 & 0 \\ 0 & -3 & 1 & 1 \end{array}\right) \to \left(\begin{array}{c|ccc} 9 & 18 & -9 & 0 \\ 0 & -5 & 3 & 0 \\ 0 & -4 & 1 & 1 \end{array}\right)$

$(18, -9, 0) + k(-5, 3, 0) + l(-4, 1, 1)$ or
$a = 18 - 5k - 4l, b = -9 + 3k + l, c = l$ for $k, l \in Z$.

(iv): $\left(\begin{array}{c|ccccc} 2 & 1 & 0 & 0 & 0 & 0 \\ -3 & 0 & 1 & 0 & 0 & 0 \\ 4 & 0 & 0 & 1 & 0 & 0 \\ -5 & 0 & 0 & 0 & 1 & 0 \\ 6 & 0 & 0 & 0 & 0 & 1 \end{array}\right) \sim \left(\begin{array}{c|ccccc} 1 & 2 & 1 & 0 & 0 & 0 \\ 0 & -3 & -2 & 0 & 0 & 0 \\ 0 & -2 & 0 & 1 & 0 & 0 \\ 0 & 4 & 1 & 0 & 1 & 0 \\ 0 & -3 & 0 & 0 & 0 & 1 \end{array}\right) \to$

$\left(\begin{array}{c|ccccc} 7 & 14 & 7 & 0 & 0 & 0 \\ 0 & -3 & -2 & 0 & 0 & 0 \\ 0 & -2 & 0 & 1 & 0 & 0 \\ 0 & 4 & 1 & 0 & 1 & 0 \\ 0 & -3 & 0 & 0 & 0 & 1 \end{array}\right).$

$(14, 7, 0, 0, 0) + k(-3, -2, 0, 0) + l(-2, 0, 1, 0, 0) + m(4, 1, 0, 1, 0) + n(-3, 0, 0, 1)$ or $r = 14 - 3k - 2l + 4m - 3n, s = 7 - 2k + m, t = l, u = m, v = n$ for $k, l, m, n \in Z$.

11:

(i): $\begin{pmatrix} 25 & 4 & | & 1 & 0 \\ -15 & 6 & | & 0 & 1 \end{pmatrix} \sim \begin{pmatrix} 5 & -16 & | & -1 & -2 \\ 0 & 42 & | & 3 & 5 \end{pmatrix} \to$

$$\begin{pmatrix} 10 & -32 & | & -2 & -4 \\ 0 & 42 & | & 3 & 5 \end{pmatrix} \to \begin{pmatrix} 10 & -32 & | & -2 & -4 \\ 0 & 84 & | & 6 & 10 \end{pmatrix}.$$

Using $k \in Z$ we cannot create $3k = 5$, therefore the system has no solution.

(iii): $\begin{pmatrix} 3 & 3 & 2 & | & 1 & 0 & 0 & 0 \\ 5 & -5 & 4 & | & 0 & 1 & 0 & 0 \\ -7 & 7 & 6 & | & 0 & 0 & 1 & 0 \\ 9 & -9 & 8 & | & 0 & 0 & 0 & 1 \end{pmatrix} \sim \begin{pmatrix} 1 & 11 & 0 & | & 2 & -1 & 0 & 0 \\ 0 & 6 & -2 & | & 1 & 3 & 0 & 0 \\ 0 & 0 & 2 & | & 0 & 23 & -1 & -12 \\ 0 & 0 & 0 & | & 0 & 55 & 2 & -29 \end{pmatrix}$

$$\to \begin{pmatrix} 2 & 22 & 0 & | & 4 & -2 & 0 & 0 \\ 0 & 6 & -2 & | & 1 & 3 & 0 & 0 \\ 0 & 0 & 2 & | & 0 & 23 & -1 & -12 \\ 0 & 0 & 0 & | & 0 & 55 & 2 & -29 \end{pmatrix}.$$

Using $k \in Z$ we cannot get $22 + 6k = 3$, so the system has no solution.

(iv): $\begin{pmatrix} 6 & 2 & | & 1 & 0 & 0 \\ -9 & -4 & | & 0 & 1 & 0 \\ 9 & 5 & | & 0 & 0 & 1 \end{pmatrix} \sim \begin{pmatrix} 3 & 2 & | & -1 & -1 & 0 \\ 0 & 1 & | & 0 & 1 & 1 \\ 0 & 0 & | & 3 & 4 & 2 \end{pmatrix} \to$

$$\begin{pmatrix} 12 & 8 & | & -4 & -4 & 0 \\ 0 & 1 & | & 0 & 1 & 1 \\ 0 & 0 & | & 3 & 4 & 2 \end{pmatrix} \to \begin{pmatrix} 12 & 8 & | & -4 & -4 & 0 \\ 0 & 5 & | & 0 & 5 & 5 \\ 0 & 0 & | & 3 & 4 & 2 \end{pmatrix}$$

$(x, y) = (-4, -4, 0) + (0, 5, 5) + k(3, 4, 2)$ or $u = -4 + 3k, v = 1 + 4k, w = 5 + 2k$ for $k \in Z$.

(v):

$$\begin{pmatrix} 3 & 4 & | & 1 & 0 & 0 \\ 4 & 5 & | & 0 & 1 & 0 \\ 5 & 6 & | & 0 & 0 & 1 \end{pmatrix} \sim \begin{pmatrix} 1 & 1 & | & -1 & 1 & 0 \\ 0 & 1 & | & 4 & -3 & 0 \\ 0 & 0 & | & 1 & -2 & 1 \end{pmatrix} \to \begin{pmatrix} 6 & 6 & | & -6 & 6 & 0 \\ 0 & 1 & | & 4 & -3 & 0 \\ 0 & 0 & | & 1 & -2 & 1 \end{pmatrix}.$$

$(x, y, z) = (-6, 6, 0) + (4, -3, 5) + k(1, -2, 1)$ or

$x = -2 + k, y = 3 - 2k, z = k$ for $k \in Z$.

References

[1] Chionglo, J. F. How I solved the linear congruence 25x = 15 (mod 29). doi:10.13140/RCR2.2.1476856324 (2016).

[2] Jones, G. A., & Jones, J. J. *Congruences.* In J. J. Jones & G. A. Jones, *Elementary Number Theory* (Springer Undergraduate Mathematics Series). London: Springer.

[3] Knuth, D. E. *The Art of Computer Programming*, Vol. 2, Chapter 4. Addison-Wesley. (1981).

[4] Medvid', V. Algorithms of solving the Diophantine equation [electronic]. In *ICERI 2023: 16th International Conference of Education, Research and Innovation, 13–15 November 2023, Seville, Spain* (conference proceedings). (2023).

[5] Mordell, L. J. *Diophantine Equations*. Pure and Applied Mathematics, Vol. 30. Academic Press. ISBN 0-12-506250-8. (1969).

[6] Shallit, J. Origin of the analysis of the Euclidean algorithm. *Historia Mathematica, 21*(4), 401–419. doi:10.1006/hmat.1994.1031 (1994).

[7] Schmidt, W. M. *Diophantine Approximations and Diophantine Equations*. Lecture Notes in Mathematics, Vol. 1467. Berlin: Springer-Verlag. ISBN 3-540-54058-X. (1991).

[8] Verdal, E. Solutions of some classes of congruences. *The Teaching of Mathematics, 9*(1), 41–44. (2006).

[9] Weinberger, P. On Euclidean rings of algebraic integers. In *Proceedings of Symposia in Pure Mathematics, Vol. 24* (pp. 321–332). Providence, RI: American Mathematical Society. doi:10.1090/pspum/024/0337902 (1973).

[10] Weisstein, E. W. Chinese Remainder Theorem. In *MathWorld – A Wolfram Resource.* Retrieved from https://mathworld.wolfram.com/ChineseRemainderTheorem.html (1999).

Index

F

G

H

I

L

N

P

R

S

T

U